Leonard Burtscher

Mid-infrared interferometry of AGN cores

Leonard Burtscher

Mid-infrared interferometry of AGN cores

VLTI/MIDI observations of 15 galaxies

Südwestdeutscher Verlag für Hochschulschriften

Impressum/Imprint (nur für Deutschland/only for Germany)
Bibliografische Information der Deutschen Nationalbibliothek: Die Deutsche Nationalbibliothek verzeichnet diese Publikation in der Deutschen Nationalbibliografie; detaillierte bibliografische Daten sind im Internet über http://dnb.d-nb.de abrufbar.

Alle in diesem Buch genannten Marken und Produktnamen unterliegen warenzeichen-, marken- oder patentrechtlichem Schutz bzw. sind Warenzeichen oder eingetragene Warenzeichen der jeweiligen Inhaber. Die Wiedergabe von Marken, Produktnamen, Gebrauchsnamen, Handelsnamen, Warenbezeichnungen u.s.w. in diesem Werk berechtigt auch ohne besondere Kennzeichnung nicht zu der Annahme, dass solche Namen im Sinne der Warenzeichen- und Markenschutzgesetzgebung als frei zu betrachten wären und daher von jedermann benutzt werden dürften.

Coverbild: www.ingimage.com

Verlag: Südwestdeutscher Verlag für Hochschulschriften GmbH & Co. KG
Heinrich-Böcking-Str. 6-8, 66121 Saarbrücken, Deutschland
Telefon +49 681 37 20 271-1, Telefax +49 681 37 20 271-0
Email: info@svh-verlag.de

Approved by: Heidelberg, Universität, Diss., 2011

Herstellung in Deutschland (siehe letzte Seite)
ISBN: 978-3-8381-3402-4

Imprint (only for USA, GB)
Bibliographic information published by the Deutsche Nationalbibliothek: The Deutsche Nationalbibliothek lists this publication in the Deutsche Nationalbibliografie; detailed bibliographic data are available in the Internet at http://dnb.d-nb.de.

Any brand names and product names mentioned in this book are subject to trademark, brand or patent protection and are trademarks or registered trademarks of their respective holders. The use of brand names, product names, common names, trade names, product descriptions etc. even without a particular marking in this works is in no way to be construed to mean that such names may be regarded as unrestricted in respect of trademark and brand protection legislation and could thus be used by anyone.

Cover image: www.ingimage.com

Publisher: Südwestdeutscher Verlag für Hochschulschriften GmbH & Co. KG
Heinrich-Böcking-Str. 6-8, 66121 Saarbrücken, Germany
Phone +49 681 37 20 271-1, Fax +49 681 37 20 271-0
Email: info@svh-verlag.de

Printed in the U.S.A.
Printed in the U.K. by (see last page)
ISBN: 978-3-8381-3402-4

Copyright © 2012 by the author and Südwestdeutscher Verlag für Hochschulschriften GmbH & Co. KG and licensors
All rights reserved. Saarbrücken 2012

To my family

Contents

1. **Introduction**   1
   1.1. Astrophysical Context 1
       1.1.1. The cosmological standard model 1
       1.1.2. The role of active galaxies in the evolution of galaxies 2
   1.2. Active Galaxies 4
   1.3. Unified models of Active Galaxies 5
   1.4. The AGN torus 8
       1.4.1. AGN torus models 8
   1.5. About this thesis 12

2. **Mid-infrared interferometry**   13
   2.1. High resolution observing methods 13
   2.2. Interferometry 14
       2.2.1. Historical Notes / Introduction 14
       2.2.2. Interferometry basics 15
       2.2.3. The (u, v) plane 20
       2.2.4. Heterodyne vs. direct interferometry 20
       2.2.5. Spatial resolution of an interferometer 21
       2.2.6. Visibility modeling with geometrical image components 22
       2.2.7. χ^2 fitting 25
   2.3. The mid-infrared wavelength region 26
   2.4. The Very Large Telescope Interferometer 29
       2.4.1. VLTI subsystems 30
       2.4.2. The MID-infrared Interferometric instrument (MIDI) 31
   2.5. Observation procedure 35
   2.6. Data reduction 41
       2.6.1. Compression of data / application of a mask 42
       2.6.2. Formation of fringes and high-pass filtering 42
       2.6.3. Determination of groupdelay 42
       2.6.4. Removal of phase biases and de-rotating groupdelay 47
       2.6.5. Coherent averaging 48
       2.6.6. Single-dish spectra 48
   2.7. Calibration and Errors 48
       2.7.1. Calibration / Visibilities vs. correlated fluxes 48
       2.7.2. Atmospheric stability / gains 50

Contents

    2.7.3. Statistical error / error budget 56
    2.7.4. Systematic errors – Repeated observations 56

3. Centaurus A: Dissecting the nuclear mid-infrared emission in a radio-galaxy   **59**
  3.1. Introduction 59
    3.1.1. A unique galaxy 59
    3.1.2. Cen A in the infrared 62
  3.2. Instrument, observations and data reduction 63
  3.3. Results 66
    3.3.1. (u, v) coverage 66
    3.3.2. Correlated fluxes 67
    3.3.3. Single-dish spectra 67
    3.3.4. Visibilities 67
    3.3.5. Variability 73
  3.4. Modelling 74
    3.4.1. Considerations for model fitting 74
    3.4.2. Geometrical models for the surface brightness distribution 75
  3.5. Discussion 87
    3.5.1. Geometrical model fits 87
    3.5.2. Variability 88
    3.5.3. The elongated source and overall geometry 92
    3.5.4. Comparison with other MIDI AGN observations 92
  3.6. Conclusions 93
  3.7. Outlook 95

4. NGC 4151: The first resolved nuclear dust in a type 1 AGN   **97**
  4.1. Introduction 97
  4.2. Instrument, observations and data reduction 99
  4.3. Results and modeling 101
    4.3.1. Single-dish spectrum 101
    4.3.2. Correlated spectra 101
    4.3.3. A possible silicate emission feature 101
    4.3.4. Simple Gaussian model 104
  4.4. Discussion 106
    4.4.1. The extended source and the Sy 1 / Sy 2 paradigm 106
    4.4.2. Greybody models and the nature of the extended source 106
    4.4.3. The point source and its relation to K band measurements 108
  4.5. Conclusions 109

5. The MIDI Large Programme: A statistical sample of resolved AGN tori   **111**
  5.1. Introduction 111
  5.2. Observations and Data Reduction 112
    5.2.1. Observational strategy 112

		5.2.2.	Target list 113

- 5.2.2. Target list 113
- 5.2.3. The observations 113
- 5.2.4. Data reduction, selection and handling 116
- 5.2.5. Uncertainties in the calibrated data 121
- 5.3. Results 124
  - 5.3.1. (u,v) coverages 124
  - 5.3.2. Correlated flux and single-dish spectra 124
  - 5.3.3. Visibilities on the (u,v) plane 124
- 5.4. Radial visibility models 127
  - 5.4.1. Results 128
    - 5.4.1.1. Other targets 148
- 5.5. Discussion 150
  - 5.5.1. Torus scaling relations 150
    - 5.5.1.1. Observational constraints 150
    - 5.5.1.2. Does distance matter? 153
  - 5.5.2. The sub-structure of tori 155
    - 5.5.2.1. Observational signs of torus substructure 155
    - 5.5.2.2. "Continuous fringe tracks" in the Large Programme 156
    - 5.5.2.3. Torus size as a function of wavelength 159
- 5.6. Conclusions 159
- 5.7. Outlook 160

6. Conclusions **163**

7. Outlook **165**
- 7.1. Developments at the VLTI 165
- 7.2. ALMA and AGNs 166

A. List of Abbreviations **169**

1. Introduction

1.1. Astrophysical Context

1.1.1. The cosmological standard model

Humans have always wanted to understand the very foundations of the cosmos we live in (Figure 1.1). And indeed, especially since the early 20th century, a lot has been learned about it: We now know that, 13.72 ± 0.12 billion years ago, the universe was in a very dense, compact and hot state, popularly known as the "big bang" (Spergel et al. 2007). A standard cosmological model emerged that, based on Einstein's Theory of General Relativity, explains the overall dynamics of the cosmos.

From exploring (among other objects) local galaxies as well as distant supernovae (Riess et al. 1998; Perlmutter et al. 1999), it was realized that only a small fraction of the universe's total energy content is ordinary, baryonic, matter. Most of the cosmos' dynamics is controlled by mysterious ingredients dubbed (cold) "Dark Matter" (CDM) and "Dark Energy" (Λ). The former is most widely seen as a real ingredient[1], provided e.g. by elementary particles that are not described within the standard model of particle physics, but can possibly be found with the current generation of particle accelerators, specifically the Large Hadron Collider. Λ is less easily explained by elementary physical theories. An attempt to identify it with the energy content of the vacuum as described by Quantum Electrodynamics, failed spectacularly by about 120 orders of magnitude. Whether it is real or, as some people believe, an account of our mis-understanding of the large-scale geometry of the universe, is an open debate.

Accepting for the time being that the nature of the main constituents of the cosmological standard model or, after its constituents, the ΛCDM model are unknown, it provides a very precise description of the dynamics of the universe. This can be used as a robust framework for simulating the formation of structure in the universe (Springel et al. 2005) (see Figure 1.2 for a timeline).

Only one of the objects studied in this work, the quasar 3C 273, is distant enough that its luminosity and distance are directly affected by cosmological model parameters. Nev-

---

[1]But there is one notable alternative theory. MOdified Newtonian Dynamics (MOND, Milgrom 1983; Bekenstein 2006) explains the dynamics of galaxies and galaxy cluster, that cannot be explained by normal matter alone and would otherwise require Dark Matter, with modified Newtonian Gravity. While recent observations (e.g. of the famous "bullet cluster", Markevitch et al. 2004) seem to disprove MOND, it nevertheless reminds us of the interesting point that we actually do not know if the fundamental laws of physics are identical all over space and time – an assumption implicitly made in nearly every astrophysical paper.

1. Introduction

Figure 1.1.: A medieval man peeks behind the appearances of the heavens and finds out how the celestial mechanics work. This image of a wood engraving by an unknown artist first appeared in Camille Flammarion's book "L'Atmosphère: Météorologie Populaire" (Paris, 1888), pp. 163

ertheless, the cosmological standard model provides the background for studying galaxy formation and evolution in which active galaxies are widely believed to play a crucial role.

1.1.2. The role of active galaxies in the evolution of galaxies

A popular scenario explaining the role of Active Galaxies in the context of galaxy evolutions was given by Di Matteo et al. (2005): During mergers, nuclear activity is triggered by the increased amount of gas that is available in the central parsecs to feed the Active Galactic Nucleus (AGN) in the center of the galaxy. But only a very small fraction of the accreted gas will end up in the super-massive black hole (SMBH) that is assumed at the center of every massive galaxy. The rest of the gas is expelled at high temperatures / velocities and reaches the rest of the galaxy where it stops star formation by increasing the temperature of the gas. Conveniently, this explains the correlation between the mass of the SMBH and properties of its host galaxy such as the bulge mass (Häring & Rix 2004) or the velocity dispersion of the stars (Gültekin et al. 2009). It also explains the observed bimodality of galaxies in color-magnitude diagrams (e.g. Skelton et al. 2009)

1.1. Astrophysical Context

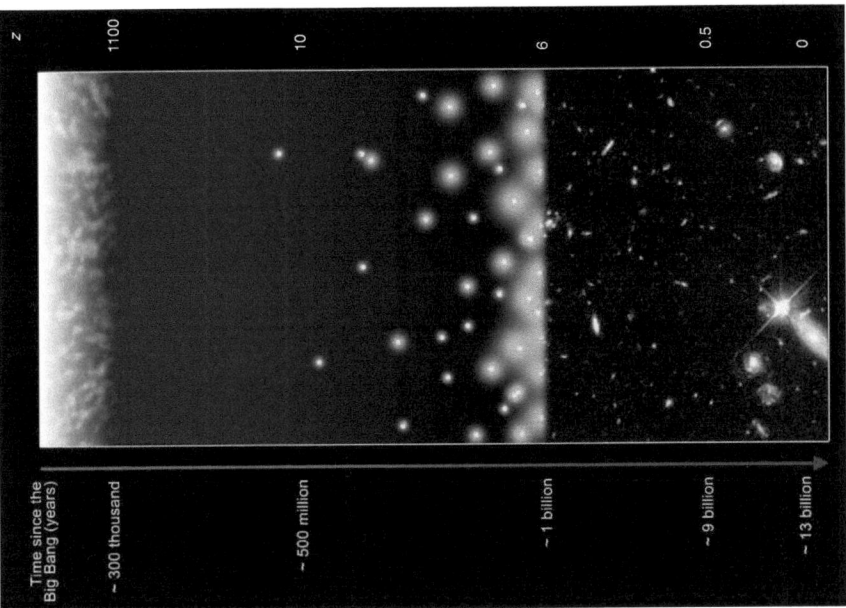

Figure 1.2.: Timeline of the universe since the big bang. After the formation of atoms out of the primordial material, the universe was ionized until it cooled down to temperatures where hydrogen could "re"-combine. From this time, about 380 000 years after the big bang, the Cosmic Microwave Background (CMB) is seen today. The time after the emission of the CMB is dubbed "The Dark Ages" since no source of light was there until the first galaxies and quasars re-ionized the universe from $z \approx 10$, a topic of intense current research. These first galaxies and quasars can today be probed with the largest ground- and space-based observatories back to about 500 million years after the big bang. The most distant galaxy observed in this work, the quasar 3C 273, is at a relatively local redshift $z = 0.158$.
Image Credit: S. G. Djorgovski et al. & Digital Media Center, Caltech; Redshifts from Ned Wright's Cosmology Calculator (Wright 2006).

1. Introduction

since the old stars in galaxies where star-formation has stopped make them appear red.

While this scheme is appealing for its explanation of several observations by this so called AGN feedback, it is only supported by numerical simulations (e.g. Croton et al. 2006). There are a number of (observational) problems with it (e.g. Robaina et al. 2009; Cisternas et al. 2011) and there is growing theoretical evidence that some of the above mentioned correlations might in fact not imply any causality (e.g. Jahnke & Maccio 2010).

While it is currently unknown what impact activity in the nucleus of a galaxy has on the evolution of its host, it is certainly true, though, that nuclear activity represents the most easily *visible* phase in the evolution of a galaxy.

In this sense active galaxies are interesting targets not only by themselves but also as a probe for studying the universe as a whole, since they are now observed out to a redshift $z = 6.4$ (Willott et al. 2007).

While quiescent galaxies, due to their larger volume density, may be detected at even larger distances – the current record holder with a spectroscopically determined redshift is at $z = 8.6$, (Lehnert et al. 2010) –, the light they emit is so feeble that detailed spectroscopic studies are often not possible (unless they happen to be magnified by a gravitational lens). Detailed spectroscopic studies of active galaxies, on the other hand, can be performed out to large redshifts allowing studies of the intergalactic medium along (and sometimes between) lines of sight to these galaxies (e.g. Hennawi et al. 2006).

1.2. Active Galaxies

Today a "zoo" of AGN types is meticulously defined by

- the luminosity in various bands,
- the existence and width of spectral lines,
- the existence or non-existence of specific features in images or maps of the galaxy and
- observation of variability.

Notable galaxy types harboring an AGN include:

- *Low Ionization Nuclear Emission Region (LINER) galaxies* are the feeblest of all AGNs, noteworthy because of their commonness: They are detectable in $\approx 50\%$ of all spiral galaxies (Ho 2008).

- *Seyfert galaxies* are more powerful than LINER galaxies but still considered lower-luminosity AGNs, hosted mostly in spiral galaxies with $M_B \geq -21.5$. This class is further subdivided in two types. Type 1 Seyfert (Sy 1) galaxies show both narrow (permitted and forbidden) and broad (only permitted) emission lines, inferring Doppler broadening by circular velocities of several hundred and several thousand (up to 10^4) km/s, respectively. The respective spatial location of these lines was termed Narrow and Broad Line Region (NLR / BLR). Type 2 Seyferts (Sy 2), on the other hand, show only the narrow lines. A further sub-classification is based on the detailed appearance of the optical spectrum and defines the so called "intermediate" Seyfert classes 1.5, 1.8 and 1.9 (Osterbrock 1981). In "diagnostic diagrams" of the strengths of various emission lines (so called "BPT diagrams", Baldwin et al. 1981), Seyfert galaxies can be clearly separated from the lower luminosity LINER galaxies and from non-nuclearly active HII (starburst) galaxies.

- *Quasars* are the high-luminosity cousins of Seyfert galaxies with $M_B < -21.5$. Historically they have also been defined to appear smaller than 7″(Peterson). About 5 – 10 % of them are strong radio sources. Due to the large luminosity of their nuclei, they appear star-like in low-resolution / low-contrast observations and they are among the most distant objects known in the universe.

- Ê*Radio galaxies* are mostly giant elliptical galaxies; some of them are quasars. They are further subdivided in Broad and Narrow Line Radio Galaxies (BLRGs / NLRGs) that can be thought of as the radio-loud analogs of type 1 and 2 Seyfert galaxies, respectively. (The designations "type 1" and "type 2" galaxies are used to generically describe galaxies with and without observed broad emission lines.) "Radio loudness" is usually defined as the ratio of luminosity at some radio frequency to the luminosity in some optical line, e.g. galaxies with $F_\nu(6\text{cm})/F_\nu(440\text{nm}) \gtrsim 10$ are called "radio-loud" (Kellermann et al. 1989).

- *Blazars* constitute the class of most rapidly variable AGNs. Their spectra show only very weak emission lines and they are always radio loud.

While this classification serves to get an overview over the variety of observed types, it is also confusing since ever more types and sub-types had to be invented as the observations revealed finer details of each galaxy. Besides that, some of the definitions are rather arbitrary and post-hoc (such as the luminosity boundary between Seyferts and Quasars), invented only to define classes, not for astrophysical insight.

1.3. Unified models of Active Galaxies

In a seminal paper, Antonucci & Miller (1985) reported that the polarized light spectrum of the Seyfert 2 galaxy NGC 1068 looks like the spectrum of a Seyfert 1 galaxy and suggested that NGC 1068 in fact has a BLR, but that our line of sight to it is obscured by dust. This gave rise to a so called "unified model" in which the two classes of (Seyfert) galaxies were explained through obscuration by a toroidal distribution of dust in the nucleus of the galaxy so that the random orientations of the galaxies with respect to our line of sight could explain the observed differences: The nuclear source (see below) is embedded in a dusty torus. Galaxies where we see the center only through this torus are called "type 2" whereas ones where we receive direct radiation from the BLR are referred

1. Introduction

Table 1.1.: Components of an AGN and their typical physical and angular sizes in NGC 1068 (distance 14.4 Mpc).

component	physical size [pc]	angular size
SMBH	10^{-5}	$1.5 \cdot 10^{-4}$ mas
Accretion disk	10^{-3}	$1.5 \cdot 10^{-2}$ mas
BLR	0.01	0.15 mas
Dust torus	2 pc	40 mas
NLR	300	6 "
Starburst	10^3	30 "

to as "type 1" (Figure 1.3). The approximate scales of this model in the nearby Seyfert 2 galaxy NGC 1068 are given in Table 1.1.

Evidence supporting this so called "line-of-sight unification" came from the observation of cone-shaped regions of ionized emission (so called "ionization cones" or extended narrow emission line regions, ENLR, for a review see Wilson 1996). They are visible in a number of active galaxies and require some confining structure. From number-counting studies of type 1 and type 2 galaxies, the observed type 1 fraction was used to determine the average value of the solid angle unobscured by the nuclear dust and was found to be roughly consistent with the ionization cone opening angles (for an estimate that avoids some of the biases affecting such studies see Maiolino & Rieke 1995).

In the context of AGNs, the term "unification" is used in the sense of finding a simple explanation for the different apparent properties of various source classes. The hope is to explain all different apparent properties with a (small) number of parameters such as viewing angle, luminosity, radio loudness etc. In the stricter sense, unification models allow only one parameter, the "viewing angle", to explain the apparent differences between the classes. Less strict versions also allow, e.g., the black hole mass and spin, the accretion rate and the radiative efficiency as parameters differing among the observed morphologies. Statistical unification models (Elitzur & Shlosman 2006), on the other hand, state that the differences between classes are determined by statistical differences in their (observed) torus properties (e.g. number of dust clumps in the line of sight).

Independently, differences in the line of sight were also made responsible for some of the observed differences in radio-loud galaxies whose radio emission arises from a jet, i.e. a narrow beam of magnetized plasma, often moving at relativistic velocities (for a review see Begelman et al. 1984). There, relativistic "beaming", i.e. the amplification of the light emitted from a relativistic source as received by an observer at rest, is employed to explain differences in spectra, appearances and variability time scales (Urry & Padovani 1995). For example the almost featureless spectra of blazars are explained this way as being almost pure beamed synchrotron radiation (which is featureless) from the jet that outshines any nuclear emission lines due to the beaming effect.

1.3. Unified models of Active Galaxies

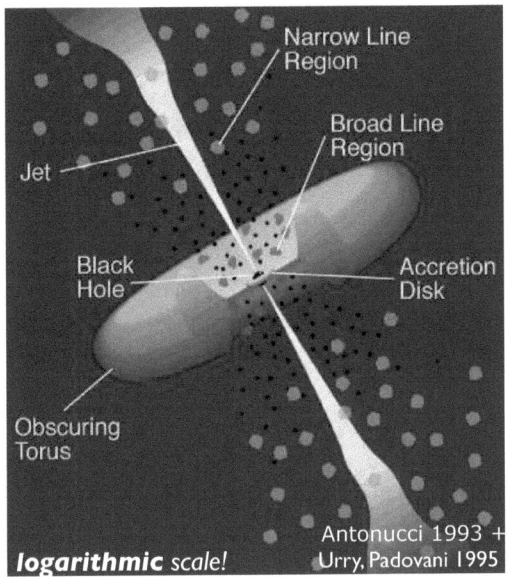

Figure 1.3.: Classical sketch of a line of sight unification model showing the main ingredients of an active galactic nucleus: An SMBH with mass M_{BH} and Schwarzschild radius $R_S = \frac{2GM}{c^2} = 3 \cdot 10^{11}$ m $\cdot M_{BH}/(10^8 M_\odot)$ is surrounded by an accretion disk (where $M_\odot$ is the mass of the sun, $M_\odot = 2 \cdot 10^{30}$ kg). Together they make up the "central engine" which is thought to be the ultimate power source of an AGN. Further out, the scenario is enshrouded by an opaque dusty torus with $R \approx 1$ pc $= 3.09 \cdot 10^{16}$ m – from some lines of sight. In the symmetry axis of the accretion disk, a relativistic jet streams outward, well collimated by magnetic fields that have been greatly amplified in the accretion process. Image Credit: NASA

1. Introduction

The "central engine": Black Hole and accretion disk At the very heart of an AGN is an accretion-powered super-massive black hole. What is an almost proven fact today, was first proposed by Lynden-Bell (1969) in order to explain the huge amount of energy being released on very small scales in AGNs. Typical luminosities exceed 10^{39} W on scales smaller than a few Astronomical Units (AUs). It was realized that such energy densities could not be provided by nuclear fusion or any other mechanism except the release of gravitational binding energy. Due to conservation of angular momentum, the in-falling matter would end up in a thin accretion disk, that is efficient in radiating away its heat (Shakura & Syunyaev 1973), or a thick low-efficiency accretion flow (ADAF, Narayan & Yi 1994), mostly depending on the accretion rate.

The theoretical limiting luminosity of the accretion process is given by the fact that the emitted radiation exerts pressure on the infalling matter. This leads to the concept of the Eddington luminosity L_E at which the inward force (gravitation) is balanced by the outward force (radiation):

$$L_E = 1.26 \cdot 10^{39} \left(\frac{M_{\mathrm{BH}}}{10^8 M_\odot} \right) \mathrm{W}. \tag{1.1}$$

1.4. The AGN torus

For a long time, there was only indirect evidence for the existence of the enshrouding dust, until Very Large Telescope Interferometer (VLTI) observations with the MID-infrared Interferometric instrument (MIDI)[2] actually resolved the parsec-scale AGN heated dust structures in NGC 1068 (Jaffe et al. 2004). Now "tori" have been observed interferometrically in a number of Seyfert 1 and 2 galaxies (Tristram et al. 2007; Raban et al. 2009; Tristram et al. 2009; Burtscher et al. 2009).

In the Circinus galaxy (Seyfert 2), an inner disk with a major axis size of ~ 0.4 pc is embedded into a similarly warm (T $\sim$ 300K), larger and almost round structure of about 2 pc in size (see Figure 1.4, left panel, from Tristram et al. 2007). Size and orientation of the inner disk perfectly agree with a rotating disk of water masers. The axis of the disk is well aligned with the axis of an ionization cone marked by [OIII] emission.

In the prototypical Seyfert 2 galaxy NGC 1068, a small (1.35 pc × 0.45 pc) disk embedded in a larger (3 pc × 4 pc) structure has also been found (see Figure 1.4, right panel, Raban et al. 2009). The small disk is hot (~ 800 K), i.e. close to the dust sublimation temperature. Unexpectedly, the axis of this disk is misaligned by $\approx 45°$ with respect to the jet axis and does also not correspond well with the ionization cone. The silicate absorption profile does not correspond to normal interstellar dust composition.

1.4.1. AGN torus models

The spatial resolution of current telescopes is not sufficient to study AGN tori and the higher-resolution interferometric observables (see the following Chapter) need to be com-

---
[2] see Section 2.4 for an explanation of the observational method

1.4. The AGN torus

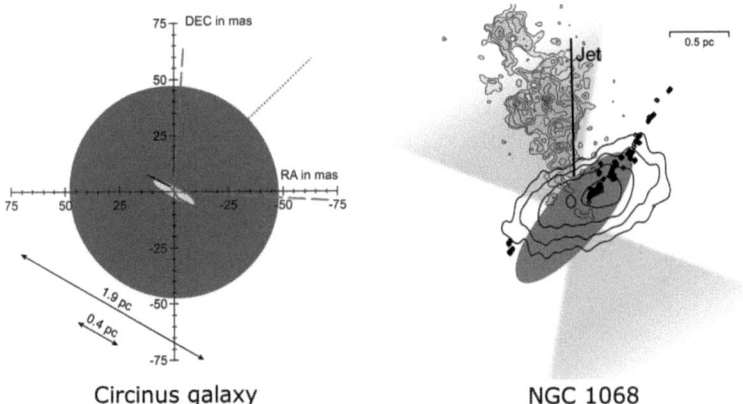

Figure 1.4.: Sketches of the nuclear dust distribution and emission of the Circinus galaxy (left, from Tristram et al. (2007)) and of NGC 1068 (right, from Raban et al. (2009)). See text for details.

pared to models to be interpreted astrophysically.

To this end, model images of AGN tori can be created by running radiative transfer simulations on model geometries, such as demonstrated by Schartmann et al. (2008) on a three-dimensional clumpy dust structure (Figure 1.5), motivated by theoretical models (Krolik & Begelman 1988).

A more physical – and more complicated – approach is the modeling not only of the appearance of tori but their actual dynamics. An example of such a fully hydrodynamical torus model was shown by Schartmann et al. (2009), Figure 1.6. This model assumes that the torus comes in place via a nuclear starburst that conveniently explains both the production of dust (required for the line-of-sight obscuration) and the release of turbulent energy required to produce a "thick" torus that is compatible with torus angles inferred from ionization cone and statistical studies.

This model has recently found observational support from a study by Davies et al. (2007) who looked for AGNs with signs of recent star formation and determined the time delay between the end of star formation and the onset of AGN activity. They find in a number of nearby galaxies that the AGN luminosity reaches values near the Eddington luminosity only after a delay of some 50 – 100 million years (Figure 1.7), confirming the basic assumption made in the model of Schartmann et al. (2009).

1. Introduction

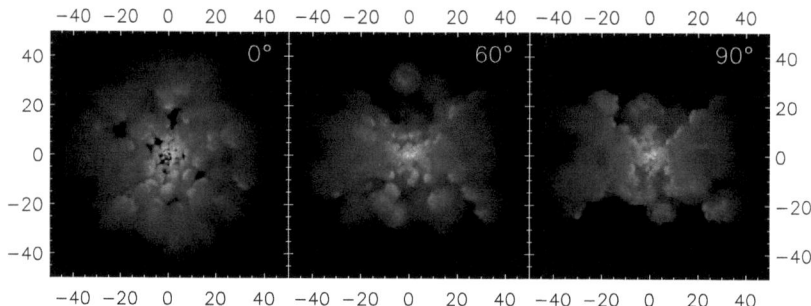

Figure 1.5.: Model images from Schartmann et al. (2008) showing the appearance of a clumpy AGN torus at 12 μm and at various inclinations.

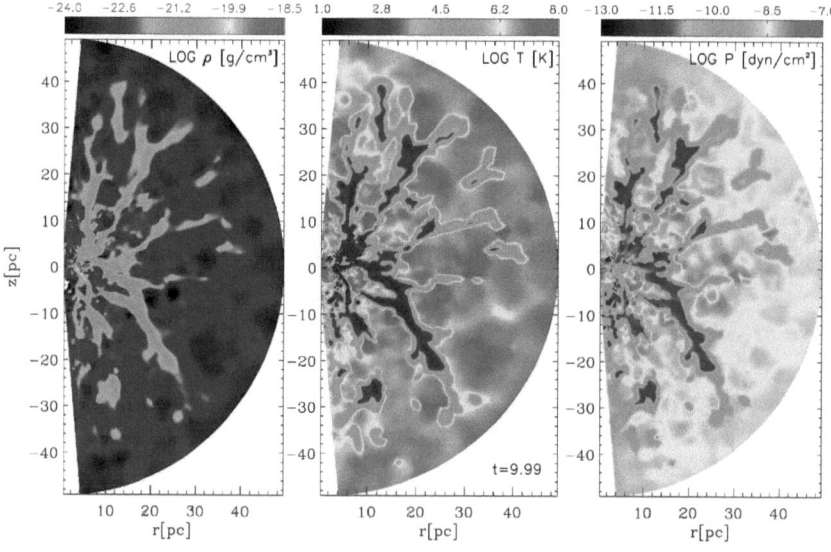

Figure 1.6.: Snapshots of the 3D hydrodynamical torus simulation by Schartmann et al. (2009) after roughly ten orbits. Left: density distribution, center: temperature distribution, right: pressure distribution. It can be seen that a filamentary structure is formed on scales of several parsecs. In the very center, a geometrically thin disk is formed, possibly connected to the disk seen in interferometric studies of nearby AGNs (see Figure 1.4).

1.4. The AGN torus

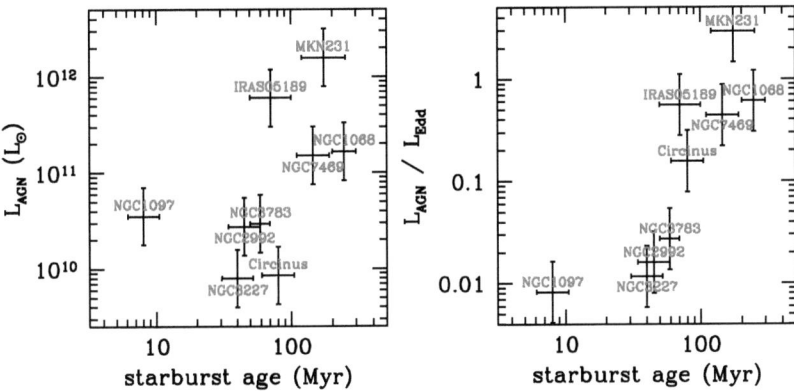

Figure 1.7.: Possible connection between the age of the most recent starburst event and the luminosity of an AGN (in units of solar luminosities, left and in units of the Eddington luminosity of the respective AGN, right), including a recent data point for NGC2992 from Friedrich et al. (2010). Figure adapted from Davies et al. (2007) by R. Davies.

1. Introduction

1.5. About this thesis

This thesis is structured as follows. In Chapter 2 the observational method is motivated and the data reduction technique is discussed. In the following three chapters, I describe the research projects on the nearby merger galaxy Centaurus A (Chapter 3), the Seyfert 1 galaxy NGC 4151 (Chapter 4) and the systematic study of a large number of AGNs in the MIDI Large Programme (Chapter 5). My conclusions are given in Chapter 6 and a brief outlook is given in Chapter 7. In the appendices, material supporting the Large Programme chapter is presented, followed by the bibliography. I profited from the help of many people, both during the course of the thesis and especially in the last months of writing and am very happy to acknowledge their support on page 167.

In the following I often use the pronoun "we" – which stands for some combination of my (senior) colleagues Klaus Meisenheimer, Walter Jaffe, Konrad Tristram and myself – to acknowledge that most of this work was motivated and many of the conclusions were discussed in this group. It does not imply that any statement has been endorsed by this group, though.

2. Mid-infrared interferometry

Interferometry is a wide field reaching from the basics of Fourier optics via the understanding of the atmospheric turbulence and instrument concepts to the intricate details and philosophies of data reduction. In this section I want to briefly present the most important concepts and basics of long-baseline stellar interferometry with a particular emphasis on those ideas that I believe newcomers to the field would like to understand but where the literature is scarce or where I found the explanations inadequate. It is not my aim to present full derivations, but essential concepts and formulae. An excellent introduction to the theory of interferometry has been given by Haniff (2007) and a more general review of optical interferometry can be found in Quirrenbach (2001).

2.1. High resolution observing methods

Even the nearest and brightest AGN tori have apparent sizes of $\lesssim$ 100 milli-arcseconds (mas) in the mid-infrared. In models, the spectral energy distribution of AGN tori has a broad peak between about 2 and $\gtrsim$ 100 μm with the prominent silicate absorption feature at $\approx$ 10 μm (e.g. Schartmann et al. 2008).

One might therefore want to use the shortest possible wavelengths to study and resolve tori since the resolving power of optical systems is inversely proportional to λ. However, tori generally are expected to appear larger at longer wavelengths (i.e. they have a radial temperature distribution). There is no simple way to determine the best trade-off between resolving power and size of the torus as its surface brightness depends on intricate details such as the possible clumpy or filamentary structure of the dust.

A number of high-resolution techniques have been developed to reach the diffraction limit of a single telescope, e.g. Speckle Interferometry (Labeyrie 1970; Weigelt 1977; Saha 1999) or Adaptive Optics (attaining the diffraction-limit in the near- and mid-infrared, e.g. Arsenault et al. 2003), or even gain a factor of $\lesssim$ 2.5 through Aperture masking (Baldwin et al. 1986; Tuthill et al. 2000, 2010)[1].

Since the Full Width at Half Maximum (FWHM) of the Point Spread Function (PSF) of an 8 m single-dish telescope in the mid-infrared ($\lambda \approx 10$ μm, see 2.3) is $\approx$ 300 mas, none of these techniques is sufficient to resolve AGN tori.

In the time-domain: Reverberation mapping An interesting way of probing very small scales in AGNs is a method called Reverberation Mapping (Blandford & McKee 1982;

---

[1] due to the selection of only the highest-spatial-frequency elements in the optical transfer function, see e.g. Hecht (2001)

2. Mid-infrared interferometry

Peterson 1993). In the spectrum of a type 1 AGN there are, among other constituents, contributions from the accretion disk very close to the BH (the optical/UV/X-Ray continuum) and from gas further out in the BLR. A variation in the continuum source (it is known to be variable in many AGNs) reaches the radius of the BLR, $R_{\mathrm{BLR}} = c\tau$, after the lag time τ. One can monitor changes in the continuum and line fluxes and cross correlate the two to get an estimate of τ. The width of the line further gives the velocity dispersion σ of the gas at R_{BLR}, so that the virial mass of the BH can be estimated to $M_{BH} \approx f R_{\mathrm{BLR}} \sigma^2 / G$ with the geometric factor f that is unknown.

This method has been applied successfully in a number of bright AGNs to derive black hole masses (Peterson et al. 2004) and even the motion of the nuclear gas (e.g. Bentz et al. 2008). It is an observationally very challenging method since it requires a dense and smooth temporal (and spectral) coverage. So far it was not possible to derive details about the spatial structure of the nuclear gas with this method.

The highest resolutions in the optical–infrared wavelength regime are nowadays achieved using optical long baseline interferometry which is discussed in the next section.

2.2. Interferometry

2.2.1. Historical Notes / Introduction

Ever since Galileo Galilei pioneered the use of a telescope for astronomical purposes in 1609, astronomers have built ever bigger telescopes to gain both sensitivity and resolution. Notable milestones in the quest for the biggest telescope were William Herschel's 40-foot (1.2 m) telescope, built in 1789, the 100-inch (2.5 m) Hooker telescope on Mt. Wilson built in 1917, famous for its use by Edwin Hubble and also for the first stellar interferometer (see below). Further notable milestones of biggest telescopes were the 200-inch (5.1 m) Hale telescope on Mt. Palomar (1949) and the 6.0 m "Big Telescope Alt-Az" (BTA) in Russia (1975). The era of the 8-10 m class telescopes has begun with the 10 m Keck I telescope in 1993.

The sensitivity of a telescope[2] with diameter D increases as D^2, the resolution only increases with D, but it is said that the cost of a telescope roughly scales with D^3.

It is therefore very expensive (and technically challenging) to obtain higher resolution images by building larger telescopes. For sources with sufficiently large surface brightnesses, a number of smaller telescopes, separated by a baseline $B \gg D$ can increase the resolution while still providing sufficient sensitivity, using the principle of interferometry.

Although the principle of stellar interferometry had been suggested by Hippolyte Fizeau already in 1868, Michelson & Pease (1921) were the first to successfully resolve the diameter of Betelgeuse with a 6 m interferometer in front of the Mt. Wilson 100-inch telescope.

---

[2] A major factor in the sensitivity calculation used to be the efficiency of the detector. But since CCDs have practically reached 100% efficiency in the visual wavelength regime, further sensitivity enhancements can only be reached by increasing the telescope diameter.

Their result of 47 mas is in good agreement with the current value of 43.56 ± 0.06 mas (Ohnaka et al. 2009). Michelson and Pease's attempts of building a larger interferometer with a baseline of 7 m did not produce any additional significant results and the further development of stellar interferometry was hindered mostly due to sensitivity and construction problems. It effectively came to a halt until it restarted with the development of the "intensity interferometer" (Brown & Twiss 1957).

Nowadays optical interferometers are an integral part of the astronomical research landscape. A non-exhaustive list of current stellar interferometers includes the Keck Interferometer (KI, Colavita & Wizinowich 2003), the Naval Prototype Optical Interferometer (NPOI, Armstrong et al. 1998), the Center for High Angular Resolution Array (CHARA, ten Brummelaar et al. 2000), the Cambridge Optical Aperture Synthesis Interferometer (COAST, Haniff et al. 2000) and the VLTI (Glindemann et al. 2000b) which I am going to discuss in Section 2.4. The Magdalena Ridge Observatory Interferometer (MROI, Creech-Eakman et al. 2010) is currently under construction; first fringes are expected for 2012.

2.2.2. Interferometry basics

At the basis of optical long-baseline interferometry is an important identity between the normalized source brightness distribution $I(\alpha, \beta)$, that we would like to observe, and the so-called spatial coherence function or normalized complex visibility $\tilde{V}(u_\lambda, v_\lambda)$, that an interferometer measures. The van Cittert Zernike theorem (e.g. Haniff 2007) states that the two are related through a Fourier transform:

$$\tilde{V}(u_\lambda, v_\lambda) = \int_\alpha \int_\beta d\alpha \, d\beta I(\alpha, \beta) \exp(-\imath 2\pi(u_\lambda \alpha + v_\lambda \beta)) \tag{2.1}$$

(α, β) are co-ordinates on the sky, e.g. (RA, DEC) and measured in radian (rad). (u_λ, v_λ) are the reciprocal co-ordinates to (α, β), pointing in the same direction ($u_\lambda \parallel$ RA and $v_\lambda \parallel$ DEC). They are *spatial frequencies* and given in fringe cycles / rad.

The normalization of the source brightness distribution is in the sense that

$$1 = \int_\alpha \int_\beta I(\alpha, \beta) \, d\alpha \, d\beta. \tag{2.2}$$

The visibilities $\tilde{V}(u_\lambda, v_\lambda)$ are normalized in the sense that $\tilde{V}(0,0) = 1$ and the amplitude of the complex visibility will be denoted as $V(u_\lambda, v_\lambda) = |\tilde{V}(u_\lambda, v_\lambda)|$.

A schematic cartoon of a two-telescope interferometer is given in Figure 2.1. The vector connecting the telescopes on the ground is $\vec{B}$. The pointing direction of the two telescopes is given by $\hat{s} = \vec{S}/|\vec{S}|$ where $\vec{S}$ is the vector from the geometrical center of the two telescopes to the distant source, $|\vec{S}| \gg |\vec{B}|$. Unless the telescopes point towards zenith, an optical delay $\hat{s} \cdot \vec{B}$ (the geometric delay) exists between the light rays reaching the two telescopes. Delay lines act to relay the light from the two telescopes to a beam combiner and to introduce the extra delays d_1 and d_2 respectively in order to correct for the geometric delay.

2. Mid-infrared interferometry

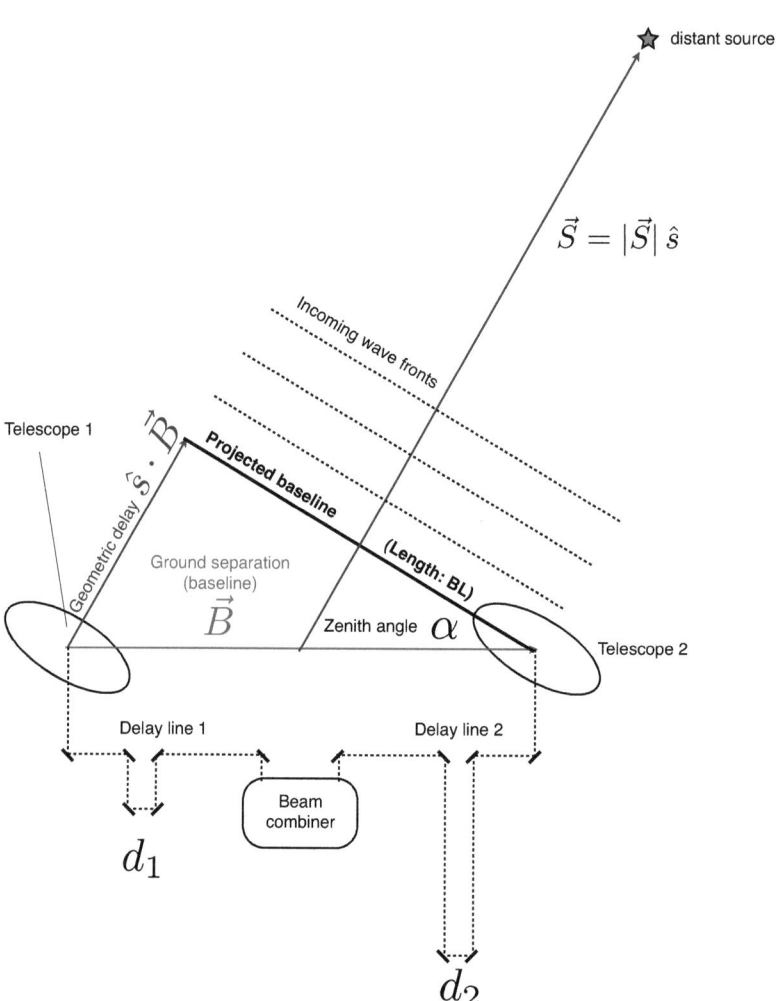

Figure 2.1.: Schematic cartoon of a two-telescope interferometer, see text for explanations.

2.2. Interferometry

The separation of the two telescopes as seen from the source defines the projected baseline vector $\vec{B}_p$ whose components (u, v) are related to the spatial frequencies (u_λ, v_λ) through

$$\vec{B}_p = \begin{pmatrix} u \\ v \end{pmatrix} = \lambda \cdot \begin{pmatrix} u_\lambda \\ v_\lambda \end{pmatrix} \tag{2.3}$$

The (u, v) co-ordinates, on the other hand, are defined by[3]

$$u = B_E \cos h - B_N \sin\phi \sin h \tag{2.4}$$
$$v = B_E \sin\delta \sin h + B_N(\sin\phi \sin\delta \cos h + \cos\phi \cos\delta), \tag{2.5}$$

where the ground separation vector $\vec{B}$ is decomposed into its components in eastern and northern direction, B_E and B_N, respectively. h is the hour angle of the source, δ the declination of the source and ϕ the latitude of the interferometer (Dyck 2000).

Further quantities of interest are the length of the projected baseline vector BL and the projected baseline vector's position angle PA (in degrees east of north). (BL, PA) and (u, v) are connected by

$$BL = \sqrt{u^2 + v^2} = |\vec{B}| \cdot \cos\alpha = \lambda \cdot BL_\lambda \tag{2.6}$$
$$PA = \tan^{-1} u/v. \tag{2.7}$$

where α is the angle between the object and zenith (see Figure 2.1).

BL is often simply referred to as "baseline length". The term "baseline" has therefore at least three meanings:

1. The ground separation vector combining the two telescopes in question, the vector quantity $\vec{B}$,

2. the projection of this as seen from the source, as in "the baseline length BL", a scalar, and

3. the name of the interferometer made up of the two telescopes (as in "the UT2–UT4 baseline").

In the following "baseline" stands for (2), the projected baseline, unless specific telescope names are given when it stands for (3), the interferometer made up of the two telescopes. $\vec{B}$ will be referred to as the "ground separation vector".

In stellar interferometers, beam combination is either performed in the image plane ("Fizeau interferometer") or in the pupil plane. The latter is called a Michelson interferometer and may be combined with fringe *detection* in the image plane (such as done in MIDI at the VLTI).

---

[3]This assumes that all telescopes have the same elevation. Otherwise additional terms need to be considered, see Dyck (2000).

2. Mid-infrared interferometry

At the beam combiner, the electric fields of the incoming waves collected at telescopes 1 and 2 with frequency $\omega/(2\pi)$ are given at time t by:

$$\Psi_1 = A\exp(\imath k[\hat{s}\cdot\vec{B}+d_1])\exp(\imath\omega t) \quad (2.8)$$
$$\Psi_2 = A\exp(\imath k[d_2])\exp(\imath\omega t) \quad (2.9)$$

where A is proportional to the collecting areas of the telescopes (assumed to be equal and set to 1 in the following) and $k=2\pi/\lambda$.

After beam combination, the detected intensity is

$$I = \langle|\Psi_1+\Psi_2|^2\rangle \propto 2 + 2\cos(k[\hat{s}\cdot\vec{B}+d_1-d_2]) \propto 2+2\cos(kd) \quad (2.10)$$

where $\langle\cdot\rangle$ denotes the time average and $d=\hat{s}\cdot\vec{B}+d_1-d_2$ introduces the optical path difference (OPD), also called "delay". In reality, d includes the atmospheric delay d_{atm} that is going to be discussed later.

This co-sinusoidal variation of intensity as a function of d is the essential observable of an interferometer: a fringe pattern. More precisely, from this fringe pattern, the Michelson visibility (also called "fringe contrast") and the phase offset at zero OPD (ZOPD) can be obtained through the maximum and minimum intensity near ZOPD ($I_{\max}, I_{\min}$) and the position of the maximum (Figure 2.2):

$$V_{\text{Michelson}} = \frac{I_{\max}-I_{\min}}{I_{\max}+I_{\min}}. \quad (2.11)$$

The crucial point here is that the Michelson visibility and the phase offset directly measure the amplitude and phase of the normalized complex visibility $\widetilde{V}$.

In practice, however, d has the additional term d_{atm} that is introduced by the atmosphere and that irrecoverably scrambles the phase information in a two-telescope interferometer[4].

So far we have implicitly assumed that our detector is monochromatic, leading to a fringe contrast $V_{\text{Michelson}}$ that is independent of d. In reality, detectors (and the atmosphere) have a finite bandpass $\Delta\lambda$ at any given wavelength λ_0 leading to a modulation of the fringes by a sinc function called the "coherence envelope" so that the detected intensity becomes

$$\begin{aligned}I' &= \int_{\lambda_0-\Delta\lambda/2}^{\lambda_0+\Delta\lambda/2} 2[1+\cos(2\pi d/\lambda)]d\lambda \\ &\propto \Delta\lambda\left[1+\frac{\sin(\pi d\Delta\lambda/\lambda_0^2)}{\pi d\Delta\lambda/\lambda_0^2}\cos(k_0 d)\right] \\ &\propto \Delta\lambda\left[1+\frac{\sin(\pi d/\Lambda_{\text{coh}})}{\pi d/\Lambda_{\text{coh}}}\cos(k_0 d)\right].\end{aligned}$$

---
[4]So called dual-beam interferometers try to 'lock' the phase of the target on a nearby reference star and can thus measure the visibility phase of the target. PRIMA is a VLTI facility that will provide phase-referencing. It is currently under commissioning.

2.2. Interferometry

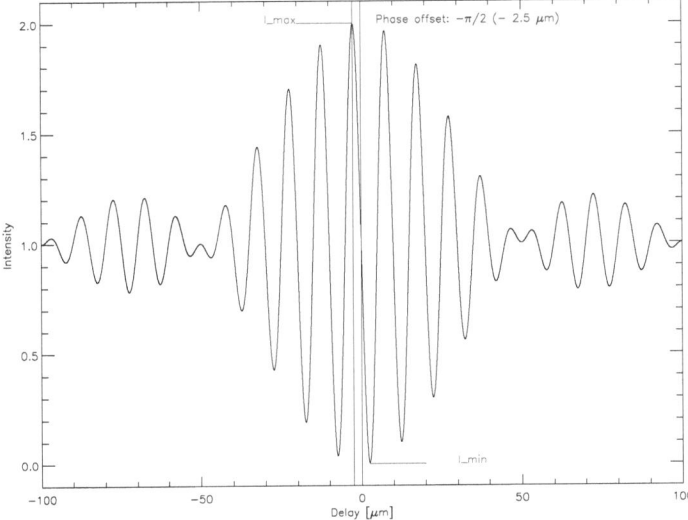

Figure 2.2.: Resulting fringe pattern for a polychromatic source with a finite bandwidth detector. In this case, the visibility amplitude is 1 and the phase offset is $-\pi/2$, corresponding to a delay offset of 2.5 μm at $\lambda_0 = 10$ μm. The spectral resolution $R = 5$ leads to a coherence length $\Lambda_{\rm coh}$ of 50 μm, recognizable by the first null of the sinc function.

where we have introduced the coherence length $\Lambda_{\rm coh} = \lambda_0^2/\Delta\lambda = R \cdot \lambda_0$ with the resolution $R = \lambda_0/\Delta\lambda$. See Haniff (2007) for a derivation.

This explains why delay lines are required at all[5]: Due to the attenuation of the fringes by the coherence envelope, d must be $\ll \Lambda_{\rm coh}$.

In order to recover the source brightness distribution $I(\alpha,\beta)$, one could imagine to simply invert Equation 2.1, given the measured values of $\widetilde{V}(u_\lambda, v_\lambda)$. This is indeed done in radio interferometry (Högbom 1974; Cornwell 2009).

In our case this is, unfortunately, not practical for two reasons: (1) In a two-telescope interferometer (without phase-referencing) only the amplitude of the complex visibility,

---

[5] The reason why *both* beams have to pass a delay line is that this is the easiest way to control for polarization equality. Only equally polarized rays interfere.

$V(u_\lambda, v_\lambda)$, is measured[6] and (2) since we will typically only be able to sample the visibility function at a few well selected (u, v) co-ordinates, the visibility function is not well defined.

Since inverting is not possible, the visibilities need to be modeled (see below, Section 2.2.6).

2.2.3. The (u, v) plane

The (u, v) co-ordinates, i.e. the components of the projected baseline vector, span the (u, v) plane (or "Fourier plane") that is best imagined as the aperture of the (huge) telescope that will be synthesized by the interferometric observations. It is used to mark the (u, v) co-ordinates of observations and serves to quickly get an overview about the observed (projected) baseline lengths and angles.

At a given wavelength, the (u, v) plane can be filled by any combination of:

- observing with various telescope combinations,
- repositioning the telescopes or
- using the earth's rotation to change the projected baseline.

The process of filling the (u, v) plane with observations is called aperture synthesis and if this is done by using the rotation of the earth it is sometimes referred to as "Earth rotation aperture synthesis".

(u, v) coverages are point symmetric with respect to the origin because the Fourier transform of a real-valued function (such as the intensity) is hermite, i.e. $\widetilde{V}(u, v) = \widetilde{V}^*(-u, -v)$. This is sometimes referred to as the "reality condition" and a less mathematical explanation for it is that the telescope positions are interchangeable.

The (u, v) coverages of the observations are displayed in the respective chapters, Figures 3.2, 4.2, 5.6.

2.2.4. Heterodyne vs. direct interferometry

The reader familiar with radio interferometry may wonder why optical interferometrists make these huge efforts (see also Section 2.4) to directly interfere the light from the two telescopes while in radio interferometry a technically much easier solution is used. In radio interferometry, the single-dish signal is mixed with a so called "local oscillator" (e.g. a maser source) of similar frequency as the signal frequency so that the latter is converted into the lower beat frequency which can be easily amplified and recorded. The recorded signals are later transferred to a correlator (a fast computer) and brought to interference there. The obvious advantage of this technique, called *heterodyne interferometry*, is that no "delay lines" need to be built and the baselines between the telescopes can therefore be very large as long as accurate clocks are available to relate the various recorded signals to

---

[6] Actually Fienup (1978) has shown that it is possible to reconstruct an image given only the amplitude of the visibility – but in *every* point of the (u, v) plane.

each other. Additionally, the sensitivity improves with number of telescopes (often called antennae) added to the array.

In direct detection, on the other hand, complicated delay lines and beam-combiners have to be built to bring the light from two or more telescopes to interference. Also, the signal is diluted when adding more telescopes to the array.

However, it is not much a matter of choice which type of interferometry one prefers, but fundamental physics sets the advantage of heterodyne vs. direct detection signal/noise rate (SNR). The Heisenberg principle, in the form $\Delta N \cdot \Delta \phi \leq 1$ (radian), states that the precision with which we can measure the phase of a photon $\Delta \phi$ is inversely proportional to the precision with which the number of photons, used for that measurement, can be determined ΔN. The number of photons in one oscillator mode is given by Bose statistics as $(\exp(h\nu/(kT))-1)^{-1}$ where T is the temperature characteristic of the received radiation. In the radio it is the so called 'antenna temperature' (usually in the range of 10-100 K) and in the mid-IR it is given by the atmosphere (that is the dominant noise source), i.e. $T \approx 300$ K. That makes $(\exp(h\nu/(kT)) - 1)^{-1}$ to be of order 10^3 at 1.4 GHz but only 10^{-3} at 10 μm and it is immediately clear that measuring the phase in the mid-IR at both telescopes individually is impossible.

The two techniques have roughly the same SNR at ≈ 100 μm; at 10 μm, direct detection provides already a ten times larger SNR over heterodyne detection and at 2 μm the advantage is $> 10^3$. See Townes (2000) for a detailed account of noise and sensitivity in interferometry.

Heterodyne detection of fringes in the mid-IR has been successfully used in a scientific instrument (ISI, Bester et al. 1990) but has been limited to the observation of very bright objects with a flux $\gtrsim 100$ Jansky (Jy). All other current infrared interferometers use direct detection.

2.2.5. Spatial resolution of an interferometer

For a single-dish telescope, the Rayleigh criterion describes two point sources of light as resolved if the center of one source's Airy disk falls onto the first minimum of the other. This leads to the well known formula for the minimum angle between two resolved sources $\Theta_{\min} = 1.22\lambda/D$ for a telescope with aperture diameter D.

In an interferometer, the angle between crests of the fringe pattern, the so-called fringe spacing, is $\Theta_{\text{fringe}} = \lambda/BL$. Applying Rayleigh's criterion to an interferometer therefore leads to $\Theta_{\min} = \lambda/2BL$.

More than in the single-telescope case, though, the resolution of an interferometer depends not only on λ and D or BL, but also on the SNR of the observation. In the single telescope case, features slightly smaller than $1.22\lambda/D$ can be discriminated if the SNR is high (and the PSF of the telescope known). In an interferometer, a source is considered to be "partially resolved" as soon as the visibility is measured to be < 1 which of course depends on the signal/noise of the observation. For realistic signal/noise values the visibility will deviate from 1 for source sizes $\ll \lambda/2BL$.

Models can therefore be discriminated at resolutions $\ll \lambda/2BL$ and in this work the

2. Mid-infrared interferometry

sensitivity to model parameters is assumed to be $\approx \lambda/3BL$ and will be called "resolution of the interferometer" henceforth. This definition of resolution is widely adopted by interferometrists (W. Jaffe, priv. comm.) and it serves our purpose. To put it in Lord Rayleigh's words: "The rule is convenient on account of its simplicity and it is sufficiently accurate in view of the necessary uncertainty as to what exactly is meant by resolution."

2.2.6. Visibility modeling with geometrical image components

The goal of visibility modeling with geometrical image components is to quantitatively model the source brightness distribution (the image) when it is not possible to directly reconstruct the image from the inverse transform of Equation 2.1.

The basic idea is to decompose the image into simple "building block" functions, such as points, (elongated) disks and rings and fit their properties (fluxes, sizes, elongations, offsets[7]) by comparing the visibility data to the Fourier Transform of the model image.

The Fourier transform of the source brightness distribution, the visibility, is generally a complex quantity. Since we cannot measure the visibility phase, however, we will restrict ourselves to point-symmetric (even) functions since the Fourier transform of a real-valued even function is real (and even).

Here the Fourier transforms of some often used components are given. The geometrical form describes the source brightness distribution.

In the following, α, β are co-ordinates in real space (in rad or mas) and the inverse co-ordinates are given as spatial frequencies u_λ, v_λ.

Most of the models are circularly symmetric and are therefore given in the radial co-ordinate $\rho = \pm\sqrt{\alpha^2 + \beta^2}$ and $BL_\lambda = \pm\sqrt{u_\lambda^2 + v_\lambda^2}$.

The models and their Fourier transforms are displayed in Figure 2.3.

The point source The intensity distribution of a point source at a position α_0, β_0 relative to the center of the interferometric field of view, may be written with the Dirac-δ function as

$$I(\alpha, \beta) = I_0 \delta(\alpha - \alpha_0, \beta - \beta_0). \tag{2.12}$$

This is the only non-point-symmetric function shown here. It can be seen that the Fourier transform of it,

$$\tilde{V}(u_\lambda, v_\lambda) = \exp(-2\pi i(u_\lambda \cdot \alpha_0 + v_\lambda \cdot \beta_0)), \tag{2.13}$$

is a complex-valued function whose amplitude is 1. This re-iterates what we have stated before: In a two-telescope interferometer, that measures only the amplitude of the visibility, it is not possible to determine the position of a point source in the interferometric field of view. It is nevertheless interesting to look also at the phase of this visibility function. It is linear in u_λ and v_λ with slope α_0, β_0, respectively, i.e. it contains the

---

[7]With phase-less data only relative astrometry within the interferometric field of view is possible, see discussion in Section 3.4.

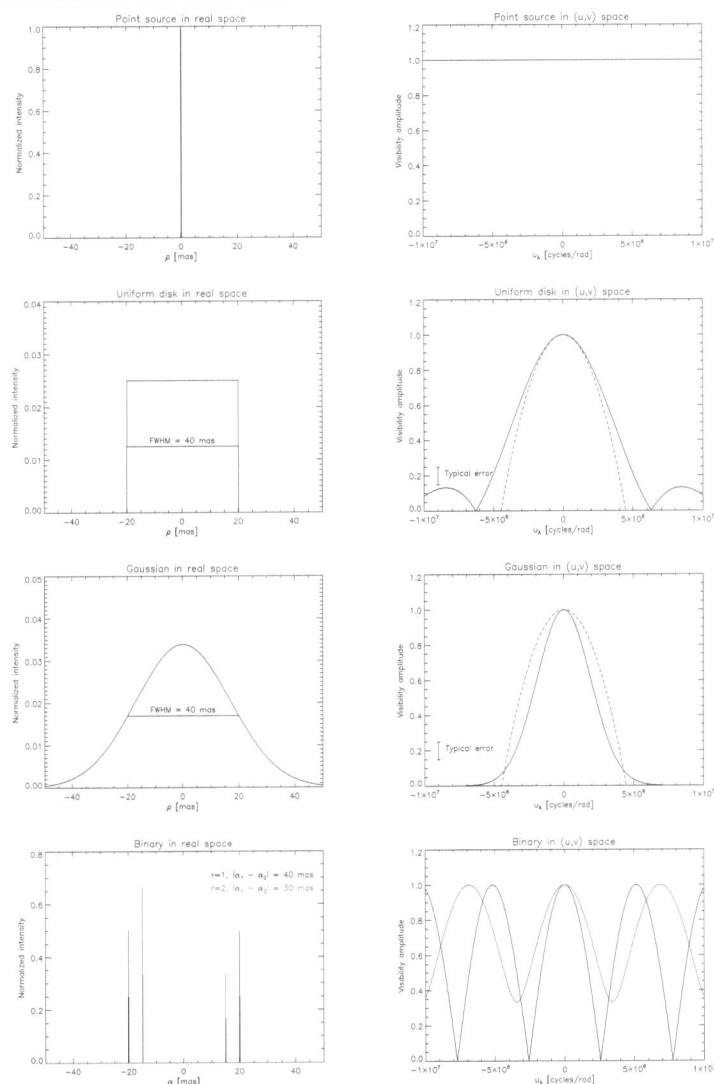

Figure 2.3.: Simple models for the source's surface brightness distribution (left panels) and their Fourier transforms (right panels). For the uniform disk and Gaussian models, a parabola of fixed width is overplotted (dashed lines) to demonstrate that the two can hardly be distinguished at the given errors (see text for details). For the binary models, the black/blue curves in the Fourier transform correspond to the respective models in real space.

2. Mid-infrared interferometry

information of *where* on the image plane a source is located. This gives an impression why some measurement of the visibility phase is crucial for reconstructing realistic images.

Uniform disk The uniform disk is the first order description of a stellar surface that is useful, for example, to model the diameter Θ of the (nearly unresolved) calibrator stars when calibrating the instrumental visibility (see Section 2.7.1). Its intensity distribution is given by

$$I(\rho) = \begin{cases} I_0 & |\rho| \leq \Theta/2 \\ 0 & |\rho| > \Theta/2 \end{cases} \quad (2.14)$$

where the normalization $I_0 = \frac{4}{\pi\Theta^2}$ (for two dimensional models) and $I_0 = 1/\Theta$ (for one dimensional models).

The visibility (amplitude) is given by

$$V(BL_\lambda) = \left| 2 \cdot \frac{J_1(\pi\Theta BL_\lambda)}{\pi\Theta BL_\lambda} \right| \quad (2.15)$$

where J_1 is the Bessel function of first kind and of integer order 1.

For highly resolved stars, a better description of the source brightness distribution takes into account limb-darkening (e.g. Hajian et al. 1998).

The circularly symmetric Gaussian disk The circularly symmetric Gaussian disk is a very popular representation of the "envelope" (i.e. the smoothed version) of a more complex source brightness distribution. It has the advantage to be smoother than the uniform disk in (u, v) space and it is therefore less likely to fit spurious signals even at moderately high resolutions. It is very useful for a first-order, reliable size estimate.

We give the equations for the Gaussian disk with zero mean both in terms of its standard deviation σ and in terms of the FWHM $\Theta = 2\sqrt{2\ln 2} \cdot \sigma \approx 2.35 \cdot \sigma$:

$$I(\rho) = I_0 \exp\left(-\frac{\rho^2}{2\sigma^2}\right) = I_0 \exp\left(-\frac{4\ln(2)\rho^2}{\Theta^2}\right) \quad (2.16)$$

$$V(BL_\lambda) = \exp(-2(\pi BL_\lambda \sigma)^2) = \exp\left(\frac{-(\pi\Theta)^2 BL_\lambda^2}{4\ln 2}\right) \quad (2.17)$$

Note how the size of the source (σ or Θ) enters the visibility function reciprocally to the real-space function, i.e., at a given BL_λ, smaller sources have higher visibilities.

The flux normalization is $I_0 = \frac{1}{2\pi\sigma^2} = \frac{4\ln 2}{\pi\Theta^2}$ (in two dimensions) and $I_0 = \frac{1}{\sigma\sqrt{2\pi}} = \frac{2\sqrt{\ln(2)}}{\Theta\sqrt{\pi}}$ (in one dimension).

In Figure 2.3, a parabola is plotted together with the Fourier transform of the uniform disk and Gaussian model (the dotted lines). It demonstrates that, at typical visibility errors ($\approx 5\%$), the three curves are actually not distinguishable (at large visibilities). Effectively it is therefore of little relevance which model is used for first-order size estimates.

2.2. Interferometry

To allow comparability between various observations it is necessary, though, to define the model that was used for the size estimate.

Binary The binary system of point sources is the simplest compound object and serves to demonstrate the visibility function of more general, complex systems. Its intensity distribution is given by

$$I(\alpha, \beta) = I_1 \cdot \delta(\alpha - \alpha_1, \beta - \beta_1) + I_2 \cdot \delta(\alpha - \alpha_2, \beta - \beta_2) \qquad (2.18)$$

and the Fourier transform is

$$V(u_\lambda, v_\lambda) = \sqrt{\frac{1 + r^2 + 2\,r\,\cos(2\pi \vec{a} \cdot \vec{\mathrm{BL}}_\lambda)}{(1+r)^2}}. \qquad (2.19)$$

The normalization is $1 = I_1 + I_2$ and the intensity ratio $r = I_1/I_2$, the separation vector of the binary is $\vec{a} = \begin{pmatrix}\alpha_1 - \alpha_2 \\ \beta_1 - \beta_2\end{pmatrix}$ and the projected baseline vector is $\vec{\mathrm{BL}}_\lambda = \begin{pmatrix}u_\lambda \\ v_\lambda\end{pmatrix}$.

The formula can easily be rewritten for the more general case of any two structures that are separated by $\vec{a}$ for which the "building block" visibility functions $V_1 = V_1(u_\lambda, v_\lambda)$, $V_2 = V_2(u_\lambda, v_\lambda)$ are known (Berger & Segransan 2007):

$$V(u_\lambda, v_\lambda) = \sqrt{\frac{V_1^2 + r^2 \cdot |V_2|^2 + 2\,r\,|V_1|\,|V_2|\,\cos(2\pi \vec{a} \cdot \vec{\mathrm{BL}}_\lambda)}{(1+r)^2}} \qquad (2.20)$$

2.2.7. χ^2 fitting

To determine the best-fitting model, a (not necessarily unique) parameter set is desired for which the deviations between the modeled visibilities, V_i^{model}, and the observed visibilities, V_i^{data} are approximately as large as expected by the observational errors σ_i.

The quadratic sum over all observations N_{data} of the so normalized deviations is given by

$$\chi^2 = \sum_i^{N_{\mathrm{data}}} \frac{\left(V_i^{\mathrm{data}} - V_i^{\mathrm{model}}\right)^2}{\sigma_i^2} \qquad (2.21)$$

The χ^2 distribution depends on the number of degrees of freedom $N_{\mathrm{free}} = N_{\mathrm{data}} - N_{\mathrm{params}}$ with N_{params}, the number of (free) model parameters. The distribution's mean value is N_{free} and its variance is $2N_{\mathrm{free}}$ (Barlow 1989). Its normalized variant is called the "reduced χ^2 function" and is given by

$$\chi_r^2 = \frac{\chi^2}{N_{\mathrm{free}}} \qquad (2.22)$$

The minimization of this quantity is a non-trivial procedure as χ^2 can be a highly complex, multi-dimensional function where the global minimum is not easily found. A

2. Mid-infrared interferometry

numerical solution to this minimization problem is provided by the Levenberg-Marquardt algorithm (Marquardt 1963).

2.3. The mid-infrared wavelength region

The wavelength band of relatively low atmospheric absorption between about 8 and 13 μm (23-37 THz) is called the N band or the "mid-infrared" region[8]. It is also referred to as thermal infrared since a blackbody of room temperature (≈ 300 K) has its maximum flux density $F_\lambda \equiv \mathrm{d}F/\mathrm{d}\lambda$ at 10 μm [9]. I will follow the astronomers' habit and mostly refer to the flux density simply as "flux" in the following.

A transmission spectrum of the earth's atmosphere (at the atmospherically privileged place of Mauna Kea, Hawaii) in this band for excellent conditions is shown in Figure 2.4 (Lord 1992). The N band is bounded by H_2O and CH_4 absorption at short wavelengths and by CO_2 absorption at long wavelengths (Cox 2000)[10]. It is dominated by a prominent double-peaked absorption feature at 9.7 μm caused by the Ozone in the earth's stratosphere. The average transmission through the feature is $\approx 40\%$ at airmass of 1 and decreases to ≈ 15 % at an airmass of 2. Due to its low transmission, the SNR in the feature is very low. It is also hard to calibrate since it varies quickly with time. Therefore we will exclude this part of the spectrum in our results.

The transmission spectrum in the N band depends strongly on water vapor. An increase in water vapor leads to a reduction in transmission and a subsequent increase in sky background emission (via Kirchhoff's law of thermal radiation). At any water content in the air, the sky background dominates almost any signal as a typical value of the sky brightness is $N \approx -3$ mag where the brightest sources studied in this work are point sources and have $N \approx 4$ mag (1 Jy). Taking into account additional background from the telescope mirrors that also shine brightly in the mid-infrared, only of order 10^{-4} of the photons detected within one PSF will actually have originated from the 1 Jy source under study.

Fortunately there are a number of ways to reduce this noise. In single-dish observations the most widely applied technique is the chopping (and nodding) of the telescope where the telescope's secondary mirror (the telescope's pointing) is switched repeatedly between the target and an off-target ("sky") position. This technique works under the assumption that the mid-infrared background varies slower than feasible chopping frequencies of a few Hz and is constant over regions larger than the extent of the relevant sources. Solid numbers about the statistics of atmospheric variations in the mid-IR are hard to find in the literature but a number of individual studies, mostly conducted in preparation for

---

[8] Actually, the mid-infrared is normally defined as the region from 5 – 25 μm, see e.g. http://www.ipac.caltech.edu/outreach/Edu/Regions/irregions.html. However, outside the N band, the earth's atmosphere makes ground-based mid-IR observations nearly impossible.

[9] Note that, in the usual representation of flux density for mid-IR spectra, $F_\nu \equiv \mathrm{d}F/\mathrm{d}\nu$, a blackbody of ≈ 500 K has its maximum flux density at 10 μm.

[10] More details about atmospheric spectral lines can be found at http://www-atm.physics.ox.ac.uk/group/mipas/atlas/.

2.3. The mid-infrared wavelength region

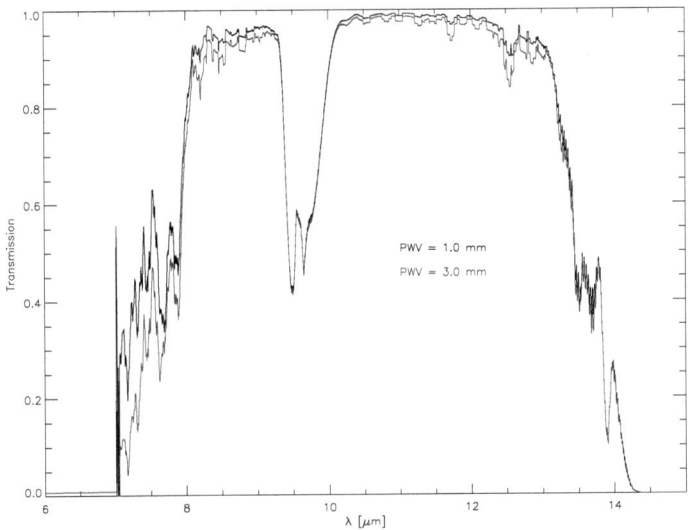

Figure 2.4.: An N band atmospheric transmission spectrum taken at the Gemini observatory (Mauna Kea, Hawaii) at airmass 1.0 and a precipitable water vapor (PWV) column of 1.0 mm (black) and 3.0 mm (blue), respectively (Lord 1992). The effect of increased water vapor is seen most clearly at short wavelengths. The spectral resolution has been downsampled to 0.2 μm to increase readability. At higher resolutions, a large number of absorption bands are seen.

a specific observing site or instrument, suggest that the minimum frequency should be several Hz and the maximum angle should be around 10″ (Kaeufl et al. 1991; Robberto & Herbst 1998). A good introduction to techniques relevant for observations in the infrared is given by Glass (1999).

Chopping and nodding can reduce the sky background by a factor of $> 10^4$ for all values of PWV observed in single-dish N band observations at Gemini (Mason et al. 2008).

At longer wavelengths, $\approx 16 - 25\,\mu$m, there is another region of atmospheric transmission, dubbed the Q band. The transmission is less than 40 % over the band and it is not widely used. For completeness, the central wavelengths of the near infrared wavelength bands are 1.3 μm (J), 1.65 μm (H), 2.2 μm (K), 3.8 μm (L), 4.5 μm (M).

2. Mid-infrared interferometry

One of the advantages of observing in the mid-IR from the ground (compared to observations at shorter wavelengths) is that observations are less affected by atmospheric turbulence since the relevant quantities scale favorably with wavelength: The characteristic length scale of atmospheric fluctuations, the Fried parameter r_0, is proportional to $\lambda^{6/5}$ and the coherence time τ is proportional to r_0/v_w with the wind speed v_w. However, the FWHM Θ of a seeing-limited PSF does not have a strong dependence on wavelength: $\Theta \propto \lambda^{-1/5}$ (e.g. Quirrenbach 2000).

Emission lines in the N band relevant for AGN research include the forbidden lines [Ar II] 8.99 μm, [S IV] 10.51 μm and [Ne II] 12.81 μm and a number of Polycyclic Aromatic Hydrocarbon (PAH) lines indicative of star forming regions (e.g. Sales et al. 2010). The most prominent spectral feature in the mid-infrared, though, is the broad absorption feature caused by vibrational resonances of the Si–O bond at $\lambda_0 \approx 9.7 \mu$m in amorphous silicates (e.g. van Boekel et al. 2005).

2.4. The Very Large Telescope Interferometer

Figure 2.5.: Aerial view of the VLT platform on the flattened mountain top of Cerro Paranal with a graphical description of how the light from the individual telescopes is guided via the light ducts to the delay line tunnel to the VLTI laboratory marked with a star. Mock-up images of three ATs are superimposed (the fourth AT was granted later). Image credit: ESO

Among the 8-10 m class telescopes, the Very Large Telescope (VLT) is unique in the sense that it actually consists of four 8.0 m so called Unit Telescopes (UTs[11]) and four 1.8 m Auxiliary Telescopes (ATs). Together they make up the VLTI (Figure 2.5, Glindemann et al. 2000a) that is arguably the most advanced and most productive[12] optical interferometer in the world.

It is located on Cerro Paranal in the Chilean Atacama desert, about 120 km south of Antofagasta and 12 km from the Pacific coast at an altitude of 2635m above sea level. The site was selected for its many clear nights, excellent median seeing and dry atmosphere (ground-layer humidity is on average 15 %). Typical current values characterizing the atmospheric turbulence at Paranal are $\Theta(N) = 0.6''$, $r_0(N) = 5.0$ m and $\tau(N) = 130$ ms (Sarazin et al. 2008).

---
[11] The primary mirror diameter is 8.2 m, but the relevant clear aperture for the input pupil is 8.0m.
[12] Roughly every fourth paper that is based on results from optical interferometry used data from VLTI, see http://apps.jmmc.fr/bibdb/ for detailed statistics.

2. Mid-infrared interferometry

Due to its latitude $\phi = -24° \, 37' \, 38''$, it offers full access to the southern hemisphere and to objects as far north as $\delta \approx +40°$. The observatory is run by the European Southern Observatory (ESO), an intergovernmental research organization financed by 14 European countries and Brazil.

The maximum ground separation of the UTs is 130 m (UT1–UT4 baseline) providing a maximum effective spatial resolution $\lambda/3$ BL = 4 mas at 8 μm. A more typical number is $\approx$ 7 mas for a (projected) baseline length of 100 m at 10 μm. For a graphic comparison: 1 mas corresponds to the size of a man seen from the earth at the distance of the moon.

The $N_{\text{Tel}} = 4$ UTs provide $N_{\text{BL}} = 1/2 \cdot (N_{\text{Tel}} \cdot (N_{\text{Tel}} - 1)) = 6$ different two-telescope combinations with ground separations ranging from 47 to 130 m. The ATs, on the other hand, can be placed on 30 different pre-defined positions, offering AT–AT ground separations from 8 to about 200m. This way several hundred (but, for technical reasons, less than N_{BL}) telescope combinations can be achieved. In principle, hybrid observations involving one or several UTs and one or several ATs are also possible, but currently not offered as a standard mode. All sources studied in this thesis are too weak to be observed with the ATs, requiring the use of the large UTs that offer about 20 times the collecting area of the ATs and supreme optics including a more sophisticated Adaptive Optics (AO) system.

2.4.1. VLTI subsystems

The VLTI's complexity is slightly underestimated as depicted in Figure 2.5. In reality, about 50 computers (S. Morel, priv. comm.) in a number of subsystems must work together faultlessly to record a fringe on the detector.

Let us follow the path of light from the first reflection to the detector: The plane wave that illuminates the apertures of the single-dish telescopes is guided through the Coudé optical trains where the AO system Multi-Application Curvature Adaptive Optics (MACAO, Arsenault et al. 2003) corrects the wavefront for atmospheric distortions. Since, in good conditions, the diameter of the Fried parameter r_0 at 10 μm is almost equal to the UT diameter, MACAO is not absolutely necessary for mid-IR interferometry as such. It is required, though, for those VLTI subsystems that operate at shorter wavelengths.

The seeing-corrected beam enters the delay line tunnel (see Figure 2.6), where the compensation for the geometrical delay $\hat{s} \cdot \vec{B}$ occurs (see Figure 2.1).

Distortions of the wavefront are not only introduced by atmospheric turbulence but also in the air-filled delay line tunnels (so called "tunnel seeing"). To reduce their deteriorating effect on the fringe detection, the Infrared Image Sensor (IRIS) detects and corrects tip–tilt wavefront aberrations in the tunnels.

After having been reflected off 16 mirrors, the beam of light reaches the VLTI lab where it undergoes another series of reflections to shape the beam and guide it into the interferometric instrument.

The VLTI is one of the most versatile interferometers in the world: it uses UTs that are normally used in single-dish mode, but it can also be used with ATs, and it combines the light of two, three or four beams in the near or mid-infrared. While this flexibility has

2.4. The Very Large Telescope Interferometer

Figure 2.6.: Panoramic view of the four VLTI delay lines in the delay line tunnel. Variable Curvature Mirrors (VCMs) and other retroreflectors ensure that the telescope pupil arrives at a fixed position in the VLTI lab (seen in the center of the image). The VCMs are mounted on carriages that glide on the delay line rails. The requirement for their position accuracy is ≈ 1 μm over the entire length of the delay lines (67m in both directions). It can be seen that there is room for more delay lines, providing an upgrade path for the VLTI to combine the light from more than four telescopes, should an appropriate interferometric instrument be built. Image credit: ESO

operational advantages, it also limits the sensitivity of the interferometer, mostly because of the large number of reflections. In the mid-infrared, for example, the throughput of the VLTI is $\approx 16\%$ and this is further reduced by the MIDI feeding optics so that only about 13% of the target flux that reaches the primary mirror is transmitted to the interferometric instrument (Puech, F. & Gitton 2006). Conversely, the *emission* of the mirrors adds to the background (noise) flux.

2.4.2. The MID-infrared Interferometric instrument (MIDI)

All observations reported in this thesis have been performed with MIDI, the adaptive-optics (MACAO) assisted interferometer in the mid-infrared for the VLTI. MIDI was built by an international collaboration led by C. Leinert at the Max Planck Institute for Astronomy in Heidelberg (Leinert et al. 2003). It was the first instrument in the world to do direct combination interferometry at 10 μm and also the first scientific instrument at the VLTI (first on-sky fringes were seen in 2002). An image of the instrument in the VLTI lab, showing mainly the warm feeding optics and the dewar on a rather compact optical bench, is shown in Figure 2.7. Selected basic parameters of the instrument are given in Table 2.1.

A schematic sketch of the instrument is shown in Fig. 2.8 and the instrument principle is briefly described there. A more detailed description of the instrument is given by Leinert et al. (2003) and a mathematical treatment of the beam combination in MIDI was given

2. Mid-infrared interferometry

Figure 2.7.: MIDI in the VLTI lab (the dimensions of the table are 150 cm x 210 cm). After having undergone a number of reflections in the VLTI lab, the beams arrive at the MIDI table (through the hole in the black structure on the left side, below which there is a white duct tape). There, they pass the so called "internal delay line" (piezo-driven roof mirrors, seen in the center of the image) and then enter MIDI through the dewar window. Figure courtesy C. Leinert

by Przygodda (2004).

Interferometric field of view The interferometric field of view (FOV) of a Michelson interferometer with finite spectral resolution $R \equiv \lambda/\Delta\lambda$ is limited to $\approx R$ resolution elements (PSFs) in diameter (Quirrenbach 2001) – it is the maximum angle at which rays interfere. Even at relatively low resolutions, this value is normally much larger than other limits to the field of view such as field stops. The VLTI field of view that is transmitted through the delay line tunnels is 2″.

Sensitivity For a target to be observable with MIDI, it needs to fulfill the sensitivity requirements of the following critical subsystems:

- **MACAO** requires a guide star with $V < 17$ mag. It can either be the source itself (e.g. all Seyfert 1 galaxies observed for this work can be guided on their nuclei)

Table 2.1.: Relevant basic parameters of MIDI and the VLTI (Leinert et al. 2003). Wavelength-dependent values are given at 10.0 μm

	VLTI
Telescope aperture D	8.0 m
UT–UT ground separations	47 ... 130 m
Resolution $\lambda/3$ BL	5.4 – 16 mas

	MIDI
Selected filters	
SiC	11.8 ± 2.5 μm
[S IV]	10.5 ± 0.2
[Ne II]	12.8 ± 0.2
Spectral resolution	$R \equiv \lambda/\Delta\lambda = 30$ (with PRISM)
Spectral resolution	$R \equiv \lambda/\Delta\lambda = 230$ (with GRISM)
(Interferometric) field of view	± 1 ″
Airy disk FWHM	315 mas
Pixel scale with field camera	$\lambda/3D \approx 100$ mas
Pixel scale with spectral camera	$\lambda/2D \approx 200$ mas (in spatial direction)
Selected slit width	200 μm (520 mas on sky)
Detector quantum efficiency	34 %

or another source of light at a maximum angular separation of 57.5″[13]. The source morphology does not matter as long as it is not too extended[14]. Its performance depends on weather conditions: The better the seeing, the brighter the star and the smaller the separation between star and science target, the better is the MACAO correction.

- **IRIS** operates in the near infrared K band. Its limiting magnitude for MIDI observations is K ≲ 11.5 mag (at ca. 1 second integration time) for so called "slow guiding" which is sufficient for MIDI operation. From experience, the detector integration time can actually be set even larger, up to ≈ 4 seconds which can be helpful for the initial alignment of weak sources that cannot be acquired using MIDI acquisition images.

---

[13]From the VLTI User Manual, http://www.eso.org/sci/facilities/paranal/telescopes/vlti/documents/VLT-MAN-ESO-15000-4552_v88.pdf

[14]Mars' polar cap ice was also used as a 'guide star' once (A. Mérand, priv. comm.)

2. Mid-infrared interferometry

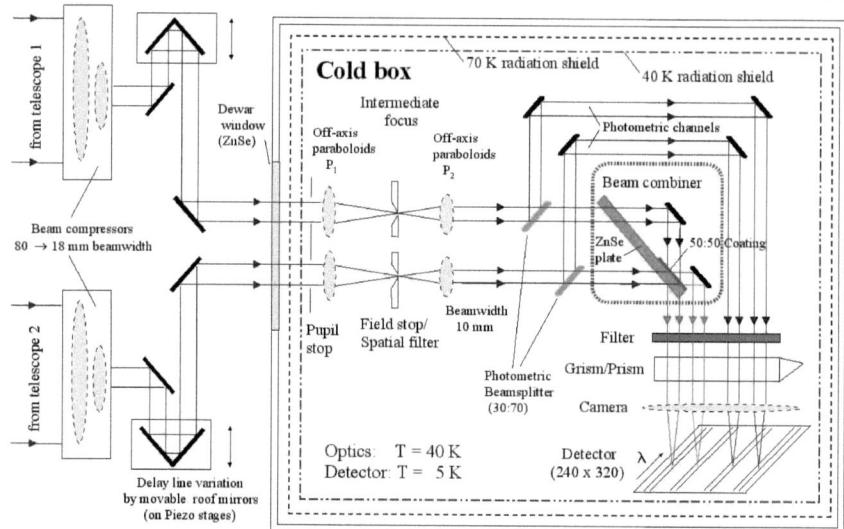

Figure 2.8.: Schematic view of MIDI: The beams pass the instrument-internal delay lines that are used as lock-in amplifiers before entering the cold optics through the dewar window. In the modes relevant to this thesis, the photometric beamsplitters are absent to gain maximum sensitivity. During interferometric observations, a 50/50 beam splitter acts as beam combiner to produce the fringe signal. It is important to note that *two* interferograms are created. Because of energy conservation, they have a phase shift of π. The signal is recorded on the detector after dispersion by a prism (the grism was not used in this work – again for maximum sensitivity). Figure courtesy C. Leinert

- **MIDI**'s official limiting sensitivity for correlated flux observations is currently given as 0.2 Jy at 12 μm in the MIDI User Manual(Rivinius, Th 2011). MIDI's ability to track a fringe depends on the atmospheric coherence time and probably also on VLTI-internal conditions, but the latter dependency is not fully understood.[15] In stable conditions, MIDI could in theory be operated without live fringe tracking using only data reduction techniques to find ZOPD. Experiments by W. Jaffe suggest that correlated fluxes as low as $\approx$ 30 mJy could be tracked that way. Offline data reduction provides a better sensitivity but requires a computer with higher performance than the MIDI control computer at the VLTI.

---

[15]Circumstantial evidence exists that the background fluctuations seen by MIDI are increased until a few days after people have entered the delay line tunnels or the VLTI lab, for example.

2.5. Observation procedure

Figure 2.9.: A typical observing scene at the VLTI console in the VLT control room. From left to right, there are two Telescope and Instrument Operators (TIOs) who control the individual telescopes, one VLTI astronomer responsible for the operation of the VLTI subsystems and one instrument scientist who, in this case, operates MIDI. To the right there is some desk space and a console for visiting astronomers.

In the following I describe an observation procedure, that has been developed by our group for observing very weak targets. It is not the standard procedure suggested by ESO and can only be executed when being personally present. For such special observing strategies, ESO offers the so called Visitor Mode that allows astronomers to personally travel to the telescope and participate in the observation. All actual instrument controls are still executed by trained personnel at the telescope, but the visitor can request settings that are not available in the otherwise offered Service Mode where observations are performed without the presence of the scientist who requested them.

Preparation Before observing with MIDI, a proposal has to be written that convinces the Observation Programme Committee (OPC) to grant time for the program (e.g. Kervella

2. Mid-infrared interferometry

& Garcia 2007). After a proposal has been accepted, the observations have to be prepared carefully (e.g. Duchêne & Duvert 2007). In order to prepare the observations one must

- collect information about the targets to be observed: search reliable co-ordinates (they must be better than the field of view), fluxes in various bands, possibly prepare a finding chart in case the target is not the only object visible within the field of view at the wavelength of observation,

- search for suitable (see below, "Calibrator" for a definition of suitable) calibrators using a calibrator database (e.g. van Boekel 2004),

- create an observing sequence taking into account the projected baseline length and angle as well as the airmass of the source,

- make sure that objects are not too close to the moon,

- for long baselines, ensure that the required delay line lengths are within specifications,

- write so called Observing Blocks (OBs – text files of a defined format that are read by the VLTI control system to set the instrument's parameters, see Figure 2.10) and

- prepare a list of backup targets in case the desired targets cannot be observed, e.g. because strong wind limits the pointing direction of the telescopes or because the targets turn out to be too weak to be observed.

A MIDI observation then consists of multiple steps involving several observing modes which have been described by Bakker et al. (2003). The ones relevant for this work are briefly introduced below.

Preset and acquisition After the sun has set, the telescope domes are opened and observing may start as soon as the target is visible above the sky background. The sky brightness in the optical does not affect mid-IR observations[16] and MIDI observations could actually start briefly after the sun has set. But since some subsystems require guide stars in the optical, one has to wait until the stars are visible in their detectors.

First, the telescopes and domes must be moved to the position of the target (the preset) and the target needs to be acquired by the various subsystems. At the same time the delay line is preset to the assumed position of ZOPD. This position is stored in a table (the "OPD model") for all telescope combinations and some pointing directions and is normally correct within a few millimeters. In order to find the exact position of ZOPD, a "fringe search" is performed (see below).

---

[16]In Rayleigh scattering, the scattered intensity at a fixed scattering angle is a strong function of wavelength, $\propto 1/\lambda^4$.

2.5. Observation procedure

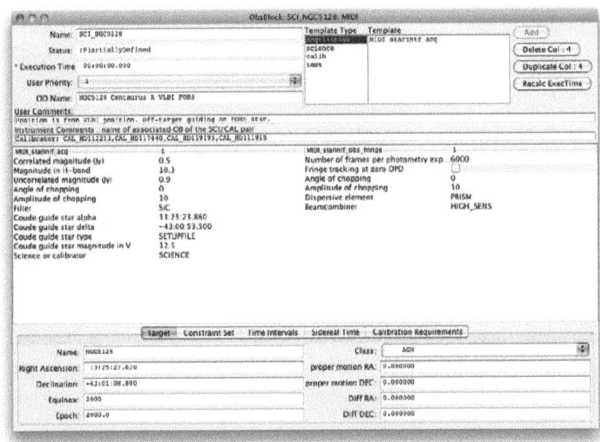

Figure 2.10.: Screenshot of an OB for the observation of Centaurus A.

The dominant overhead in this step is the acquisition of the MACAO guide star. The reason for this is that MACAO has very sensitive detectors[17] that require a careful selection of neutral filters to ensure maximum sensitivity while not damaging the detector. This is done by switching in and out numerous filter combinations and takes time.

After the source is acquired in MACAO, the source is centered and tracked in the VLTI lab using IRIS ("lab guiding"). This step only takes a few seconds. Traditionally the source would next be acquired in MIDI by taking images of the source on the MIDI detector with one of the filters listed in Table 2.1. The beam splitter is removed in acquisition to get separate images of the two beams. For the purpose of centering the source on the MIDI detector, it suffices to check the relative alignment between IRIS and MIDI once or twice in the night and otherwise rely on the "lab guiding". For weak targets, acquisition images may still be taken with MIDI in order to have an additional dataset to search for systematic errors. However, to my knowledge, the acquisition images have never actually been successfully used for that purpose.

Already evident from the acquisition images is a large thermal background that is not fully removed through chopping (Figure 2.11, right panel). In Figure 2.12 a time sequence of the acquisition images is shown, demonstrating that (1) it is nearly impossible to center very weak sources with acquisition images in MIDI (and the need for a good IRIS acquisition) and, more importantly, (2) that the background is very uneven. This is the same background that will affect the single-dish spectra.

The exact cause of these background fluctuations is unknown. A number of factors has

---

[17] Avalanche Photodiodes

2. Mid-infrared interferometry

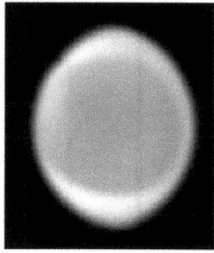  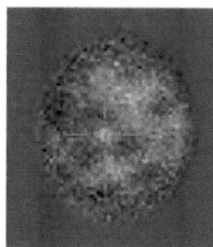

Figure 2.11.: Acquisition images taken with MIDI in the night of 2010-05-27 (only the image of one of the two beams is shown). From left to right: Sky with $\approx$ 40000 counts/s (the central round area of 2″ diameter is the sky; the bright edge around it is the warm delay line tunnel), a calibrator with barely visible Airy ring (500 counts/s) – showing that MACAO produces diffraction limited imaging in the mid-IR – and a weak target (NGC 4593, 15000 frames, 5 counts/s)

been suggested that could cause such variations but no convincing proof has been given as to which of them is the most relevant one. These factors include

- atmospheric variations,
- errors in the wavefront correction by the MACAO,
- errors in the tip–tilt correction by IRIS and
- possibly uneven surfaces in the VCMs.

A more thorough investigation is needed in case single-dish spectra of weak sources are a desired observing mode for MIDI.

Fringe Search Once overlap of the beams from the two telescopes is produced (either by alignment in IRIS or in MIDI), the fringe search can begin, either in "slow" or "fast" mode.

If the source is bright enough, the higher spectral resolution grism (see Table 2.1) is used, providing a larger coherence length $\Lambda_0 = R \cdot \lambda$ (Equation 2.12), thereby making it possible to use larger step sizes (500 μm) in the search for the zero OPD position than with the prism. The fringe search process is necessary since the OPD model is not accurate enough to preset the delay lines within one coherence length of the zero fringe position.

Fringe Track Once the fringe is found[18] (for a certain telescope combination and at a certain pointing direction of the telescopes), the actual fringe integration can begin.

---
[18]"The fringe" is jargon for the ZOPD position.

2.5. Observation procedure

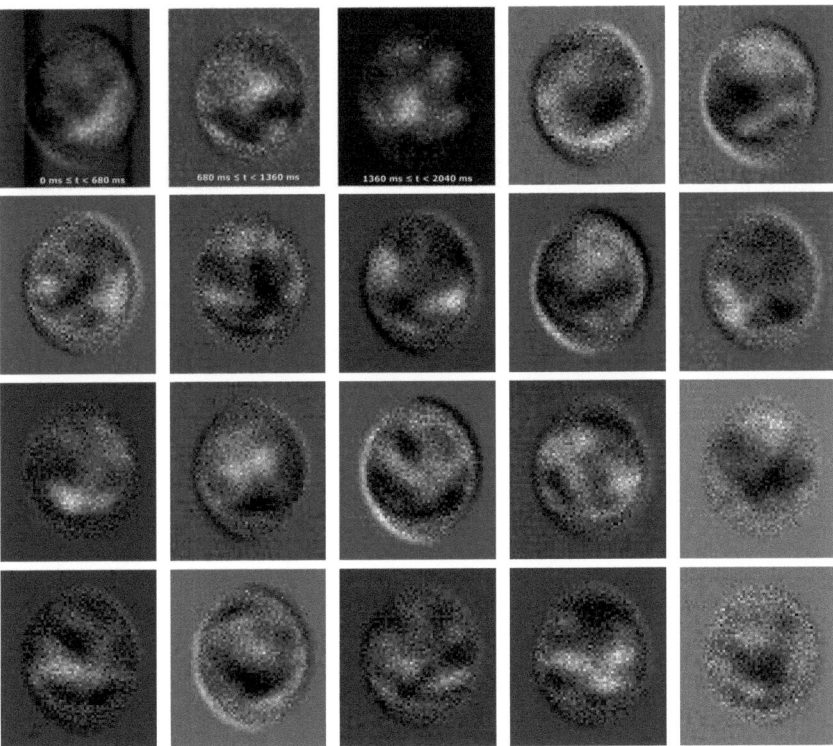

Figure 2.12.: Acquisition image sequence for NGC 4593 (from top left to bottom right, row by row). Each image is composed of 200 target and 200 off-target frames, i.e. shows a snapshot of 680 ms (the Detector Integration time, DIT, was 1.7 ms). Near the center of the image the source is visible in some snapshots, but other bright spots appear and disappear, demonstrating that the background is very uneven and varies fast in time. The chopping rate was 1.5 Hz and the summed image has been shown in Figure 2.11 (right panel)

2. Mid-infrared interferometry

In this step, the VLTI delay lines (blindly) follow the *expected* fringe position by compensating for sidereal motion according to the OPD model while the atmospheric delay is determined by the MIDI-self fringe tracker and corrections are sent to the main delay lines. During that time the MIDI internal delay lines move in a sawtooth pattern of amplitude 80 μm. This way a known modulation is added to the fringe signal that can be very effectively used to remove the background noise. Per default, each scan consists of 40 steps of 2 μm each and typically 200 such scans are performed for weak sources, leading to 8000 frames. The maximum integration time is set by the requirement not to saturate the detector and is usually chosen to be 18 ms.

The atmospheric OPD can be seen as a random walk process with a ≈ 1 μm step per second. In order to compensate for it, the position of ZOPD is determined from the data in nearly real time in almost the same way as described for the data reduction in Section 2.6.3.[19]

In fact, the fringe can be sampled either around ZOPD value (i.e. the tracking OPD crosses ZOPD) or around an offset value of 50 μm (i.e. the tracking OPD varies between +10 and +90 μm). The latter has some advantages for background subtraction when observing very weak sources and is explained in Section 2.6.3.

Photometry The last step of a MIDI observation of a target is the measurement of the single dish spectrum, called the "photometry". It is done sequentially by first blocking the light from one telescope and recording the signal from the other, then vice versa. Otherwise, the optical setup of MIDI is identical as for the fringe measurement, in particular the flux of each beam is measured through the beam combiner.

During photometry, the secondary mirrors of the telescopes are chopping. For our observations we choose a chopping frequency of 1.5 Hz and a chopping throw of 10″. While a higher chopping frequency would be desirable to better reduce the mid-infrared background, it would on the other hand reduce the number of usable frames drastically. The reason for this is that a fixed number of frames for each change of chopping position (target / sky) must be discarded due to the time it takes for MACAO to re-acquire the AO lock.

Calibrator The observation sequence is then repeated for a nearby calibrator star with known mid-infrared flux in order to determine the atmospheric and instrumental transfer functions.

The ideal calibrator star is

- unresolved,
- has the same flux as the target source and

---

[19]This method of fringe tracking is referred to as "self-fringe tracking" since the tracking is done on the science data themselves. An alternative is external fringe tracking. However, the external fringe trackers at the VLTI that work(ed) together with MIDI – FINITO and PRIMA – were either not sensitive enough (FINITO) or are not yet available for open time observations (PRIMA). The online self-fringe tracking system does not include the "weak source routines", described below.

- has minimum separation on the sky from the target, especially in airmass.

The actual calibration that is performed in the data reduction is described in Section 2.7.

2.6. Data reduction

In order to derive scientifically useful spectra from the raw data recorded at the telescope, the interferometric data reduction software *MIDI Interactive Analysis + Expert Work Station* (MIA+EWS Jaffe et al. 2004; Köhler & Jaffe 2008)[20] was used. It consists of the two parts

- *MIA* that employs the so called incoherent data reduction and has a user-friendly graphical user interface and

- *EWS* that uses coherent data reduction and is accessible via IDL[21] scripts and C programs.

For weak sources with $F_\nu(12\mu m) \lesssim 1$ Jy, only the coherent data reduction approach leads to sufficiently high SNR results, which is why only EWS was used for the reduction of all data described in this thesis. For extremely weak sources with $F_\nu(12\mu m) \lesssim 0.5$ Jy, the standard data reduction by EWS, as described in detail by Tristram (2007), can be further improved. An ongoing development process, led by W. Jaffe, has been started to adapt EWS for the reduction of extremely weak targets.

It has not yet led to a stable release version, so the reference for the following data reduction description is the nightly built "snapshot version" of MIA+EWS of 2 February 2011.

The following description of the data reduction is partly based on and complementary to the description given by Tristram (2007).

The most recent additions to MIA+EWS are also discussed on the MIDI wiki[22]. The mathematical foundation for the coherent data reduction has been described by Jaffe et al. (2004) and Meisner et al. (2004).

The main difficulty in reducing mid-IR interferometric data of very weak targets is the extremely low SNR in the raw data. The background count rate exceed the signal count rate by $\gtrsim 10^4$ for sources as week as $\lesssim 0.5$ Jy. This is the main driver of all the steps described below.

---

[20]The software is released under an open-source license and can be downloaded from http://www.strw.leidenuniv.nl/~jaffe/ews/

[21]Interactive Data Language, a proprietary scientific programming language by ITTVIS, http://www.ittvis.com

[22]http://www.mpia-hd.mpg.de/MIDISOFT/wiki/

2. Mid-infrared interferometry

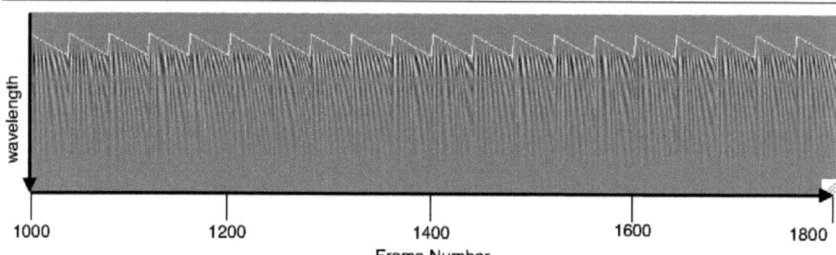

Figure 2.13.: Fringes of a calibrator star after high pass filtering, showing the position of the MIDI internal delay line on top.

2.6.1. Compression of data / application of a mask

The first step in the data reduction is the application of a spatial filter (a mask) to transform the two-dimensional detector image into a one-dimensional spectrum.

The mask should ideally have the same shape as the signal to maximize the SNR.

2.6.2. Formation of fringes and high-pass filtering

In this step, the two interferograms (cf. Figure 2.8) are subtracted from each other. The interferometric signal from the two telescopes is correlated and has a phase shift of π. The noise is uncorrelated between the two telescopes (they see different patches of sky). A subtraction of the two interferograms reduces the noise by $\approx 90\%$.

A further, even more efficient noise subtraction is reached by high-pass filtering the data. During the fringe track process, the fringes were modulated by the MIDI-internal delay lines. Smoothing the one-dimensional data in time-direction therefore reveals only noise that we can subtract from the data. This further reduces the noise by a factor of ≈ 100 (W. Jaffe, priv. comm.).

For this work the data have been smoothed by a box-car filter with a width of 10 frames. After this processing step the fringes are visible in the data of a bright source (Figure 2.13).

2.6.3. Determination of groupdelay

The next steps of the data reduction process, described in this paragraph, are necessary to determine the atmospheric delay d_{atm}. Let us write the recorded intensity (in counts/s) as

$$I = I_{\mathrm{src}} \cdot V(BL_\lambda) \cdot \sin(kd + \phi), \qquad (2.23)$$

where I_{src} is a term that is proportional to the intensity of the source and is measured in counts/s and k is the wavenumber $2\pi/\lambda$. Since the delay lines have removed the geometric delay (Figure 2.1), the delay d entering the above equation is

2.6. Data reduction

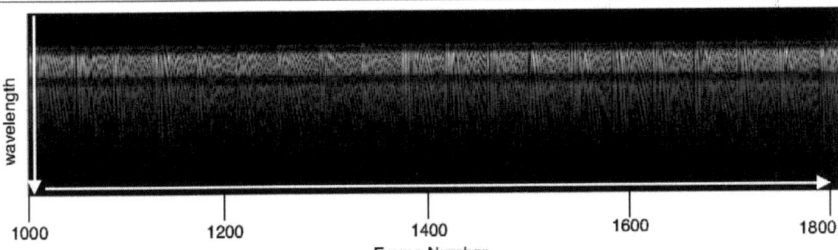

Figure 2.14.: Fringes after removal of instrumental delay d_ins. Here the absolute value of the complex quantity I_rot is shown.

$$d = d_\text{ins} + d_\text{atm} = d_\text{MIDI} + d_\text{VLTI} + d_\text{atm}, \qquad (2.24)$$

where d_ins is the instrumental delay, composed of d_MIDI, the delay-modulation introduced by the MIDI-internal delay lines and d_VLTI, the compensation for the atmospheric delay that the online fringe tracking system applied while observing.

If the fringe tracking had been perfect, it would have fully compensated for the atmospheric delay (i.e. $d_\text{VLTI} = -d_\text{atm}$) and we would only need to take care of the (known) contribution d_MIDI before coherently averaging all data. Unfortunately this is not the case.

In the next step of data reduction, the delay, or more precisely, the groupdelay $\tau_g(\nu)$ is therefore determined from the data. The groupdelay is defined as

$$\tau_g = \frac{1}{2\pi} \frac{d\phi}{d\nu}, \qquad (2.25)$$

where $\nu = ck/(2\pi)$ is the frequency of the signal.

In the next step, the instrumental delay is removed from the data by "rotating" the data by $\exp(-\imath k d_\text{ins})$. The data then become complex and can be written as

$$I_\text{rot} = I \cdot e^{-\imath k d_\text{ins}} \qquad (2.26)$$

$$= I_\text{src} V(BL_\lambda) \frac{1}{2\imath} \left(e^{\imath k d_\text{atm} + \imath \phi} - e^{-\imath k (d_\text{atm} + 2 d_\text{ins}) - \imath \phi} \right). \qquad (2.27)$$

The absolute value of I_rot is displayed in Figure 2.14.

Since we want to determine the atmospheric delay, we perform a Fourier transformation on I_rot. The result,

$$F(d) = \int I_\text{rot} \cdot e^{-\imath k d} dk, \qquad (2.28)$$

is called the *delay function* and is displayed in Figure 2.15 for a weak target. The top most panel of that Figure shows three peaks per column (frame):

43

2. Mid-infrared interferometry

1. One that moves very fast with frame number: This corresponds to the second exponential function (the one with $2d_{\text{ins}}$) of Equation 2.27.

2. One that moves with frame number, but not as fast as the first one. This is the (uncorrelated) sky background that remains even after the high-pass filtering, done in the first step. Since we multiplied the whole signal by $\exp(-ikd_{\text{ins}})$, it now changes its phase with frame number. Its spectrum is that of a blackbody, i.e. it is almost flat over the N band – like the real signal at ZOPD! By tracking the source not at ZOPD but at a ZOPD-offset of a few times λ, the true signal becomes modulated in k-space (it varies quickly with wavelength) and this information can be used to suppress the sky signal (this option is called "dave" in EWS). This suppresses the noise by another factor of ≈ 10 and is shown in Figure 2.15 in the second panel from top.

3. One that moves only very slowly with frame number. This is the atmospheric delay that we wish to obtain.

To increase SNR and to remove the fast varying peak, a Gaussian smoothing process with standard deviation σ_{gsmooth} is applied. This smoothing process is demonstrated in the two lower panels of Figure 2.15 for two values of σ_{gsmooth}.

In the standard EWS routine, the groupdelay for each frame is determined from this data as the maximum value of the delay function (i.e. where the brightest spot in each column of the image is). For bright sources the peak shows up very clearly in the image, but for weak sources (like the one shown in Figure 2.15 with ≈ 200 mJy correlated flux) where the SNR is not much larger than unity, it can be hard to find the signal peak among the many noise peaks. The so found groupdelay is displayed in Figure 2.16. It can be seen that this procedure works very well for bright sources but is not very efficient in finding the correct groupdelay for weak sources since the algorithm is very often confused by the noise and chooses a wrong value of the delay function.

Values of the groupdelay that differ a lot from the previous value or that are excessively large compared to the average groupdelay are very probably noise peaks and automatically rejected by a flagging routine so that only "good" frames, i.e. those for which the groupdelay is well determined, are kept for the averaging process later.

From a comparison between the groupdelay found for the bright calibrator target (Figure 2.16, upper panel) and the one found for the weak source target (Figure 2.16, lower panel) it can also be seen that this method does not make optimal use of our previous knowledge, namely: that the atmospheric delay does not change very much (in not too bad nights)[23]. This information is used for the weak sources routines.

---

[23]The change of atmospheric delay with time is described in the context of Kolmogorov turbulence (e.g. Quirrenbach 2000). The relevant result for the groupdelay estimation is that the change of atmospheric delay with time can be estimated to be $\Delta d_{\text{atm}} \approx 1$ μm/s $\cdot t$ within a certain range $t_{\text{min}} < t < t_{\text{max}}$. The lower time limit ($t_{\text{min}}$, high-frequency limit) is given by the fact that the telescopes have a finite diameter and therefore changes due to moving convection cells cannot occur instantly. The upper limit (t_{max}) occurs because the ground separation of the telescopes is finite.

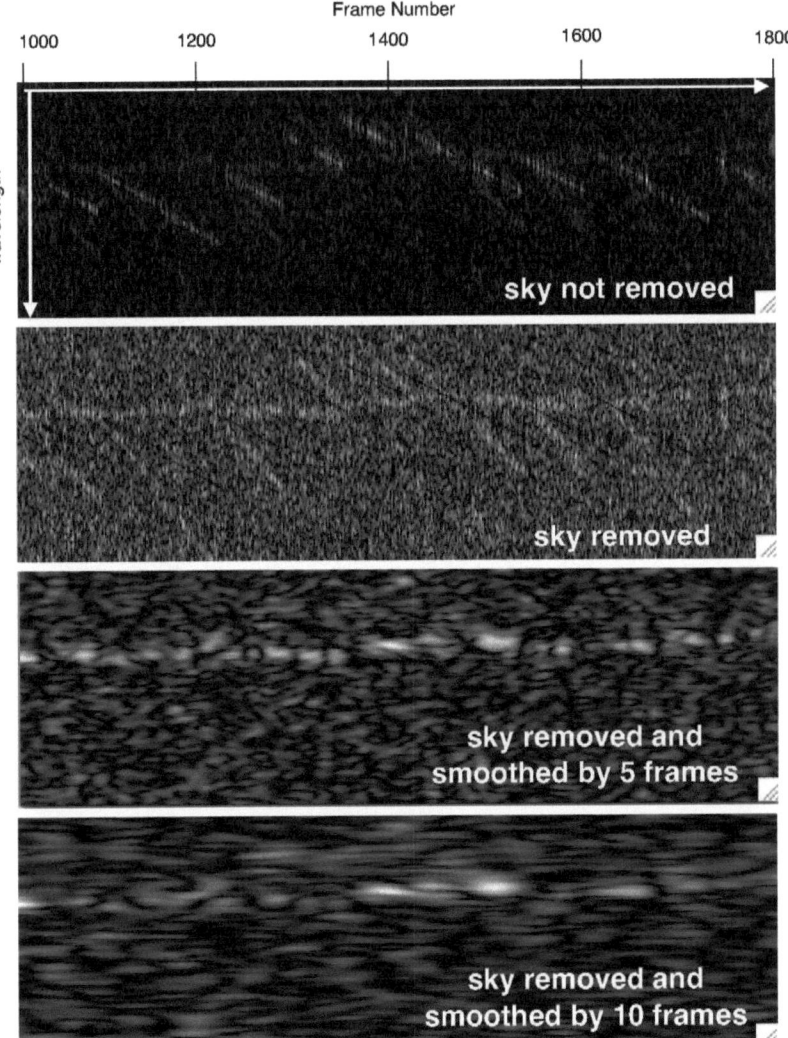

Figure 2.15.: Top panel: A very noisy delay function (shown is the amplitude of the complex quantity $F(d)$ (Equation 2.28) for the weak science target NGC 4151 showing three peaks per column (see text for details). Second from top: The same delay function but now with the sky signal removed, two peaks per column remain. Second from bottom and bottom: smoothed (and oversmoothed) versions of the delay function with $\sigma_{\text{gsmooth}} = 4 \cdot 18$ ms and $10 \cdot 18$ ms, respectively.

2. Mid-infrared interferometry

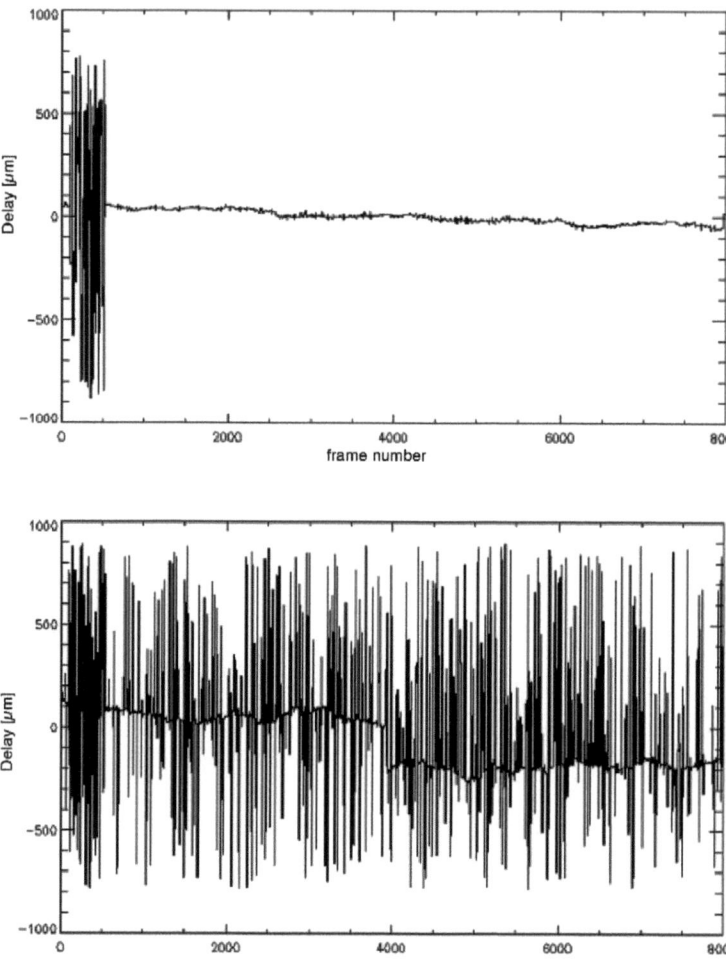

Figure 2.16.: Top panel: Groupdelay determined for a bright calibrator star. At the beginning of each fringe track, the delay line is deliberately moved to a position far off the fringe ($d \gg \Lambda_{\mathrm{coh}}$) to determine the noise level. During this time (first ≈ 600 frames), the groupdelay cannot be determined. Bottom panel: Groupdelay determined for a very weak science target. Apart from noise, a delay line jump (a so called "glitch") is seen shortly before frame number 4000. Both cases show a smoothing $\sigma_{\mathrm{gsmooth}} = 4 \cdot 18\mathrm{ms}$.

2.6. Data reduction

Modification of groupdelay estimation for weak sources:

- In order to improve on the groupdelay finding, one can run the groupdelay search in two steps (this technique is called *slow fitting*):
    1. In the first approximation, the true groupdelay is found by averaging[24] the Fourier transformed images (Figure 2.15, top panel) by a large time (≈ 10 seconds). This removes most of the noise, but returns a delay function that is only accurate to $\approx 10~\mu m$ (the OPD varies by about 1 μm per second in average conditions).
    2. With this delay approximation, another groupdelay search is run, now with a narrower averaging width, and only looking in an envelope of $\approx 10~\mu m$ around the previously found estimate. This is a reasonable way of finding the groupdelay of weak sources under not too bad conditions where the true atmospheric OPD does not change by more than $\approx 1~\mu m$ per second.

- An extra way of improving the signal to noise in groupdelay searches for coherent integration is the so called *slope fitting*. The value of σ_{gsmooth} that can be applied to the *complex* image is limited by the atmospheric coherence time. But if we consider a linear drift and *try* various changes of OPD within the time defined by σ_{gsmooth} (and choose that change that leads to the maximum signal), we can allow σ_{gsmooth} to be larger by up to a factor of two, improving the SNR of the groupdelay finding algorithm by $\sqrt{2}$.

2.6.4. Removal of phase biases and de-rotating groupdelay

Now that the atmospheric delay is known, a second rotation is performed to co-phase all frames:

$$I_{\text{rot2}} = I \cdot e^{-\imath k d} \qquad (2.29)$$

$$= I_{\text{src}} \cdot V(BL_\lambda) \frac{1}{2\imath} \left(e^{\imath \phi} - e^{-\imath(2kd+\phi)} \right) \qquad (2.30)$$

The removal of the groupdelay is effectively a fit to the gradient of the phase with respect to frequency (see Equation 2.25 and Figure 8b of Jaffe 2004). After removing it, a linear component (an offset) of the phase remains that does not depend strongly on wavelength, but changes with time and must be removed before coherently averaging the data.[25]

---

[24]The groupdelay is uncorrelated between frames that are separated by more than the atmospheric coherence time τ_0, i.e. their phase relation is random. Averaging must therefore be performed on a real quantity, in this case the amplitude of the complex delay function.

[25]The removal is done by averaging the offset over all wavelengths per frame and removing this average. It is important to do this averaging process per wavelength and *exclude* the wavelength for which the phase offset is being determined in the average. Otherwise a bias is introduced.

2. Mid-infrared interferometry

The reason for the appearance of this offset is that the geometrical delay (that occurs in vacuum / in space) is compensated for in air-filled delay lines. Mostly the water vapor there introduces this phase shift (dispersion) ϕ_{H_2O}.

2.6.5. Coherent averaging

Now that the groupdelay and phase shifts are removed we can simply sum up all N frames to give the final result

$$\begin{align}
I_{\text{coh}} &= 1/N \int_t I_{\text{rot2}} \cdot e^{-i\phi_{H_2O}} dt \tag{2.31} \\
&= 1/N \int_t I_{\text{src}} \cdot V(BL_\lambda) \frac{1}{2i} \left(e^{i\phi_{\text{diff}}} - e^{-i(2kd+\phi-\phi_{\text{diff}})} \right) dt \tag{2.32} \\
&= I_{\text{corr}}(BL_\lambda) \cdot e^{i\phi_{\text{diff}}} \tag{2.33}
\end{align}$$

During the data reduction process we removed the offset and the linear slope in phase (i.e. the groupdelay). The phase that remains is the so called *differential phase* $\phi_{\text{diff}} = \phi - \phi_{H_2O}$.

It contains important, but hard to interpret, information about the source and can be used for modelling. So far this has never been done for MIDI data of weak sources. The result that we will be using in the following chapters is the amplitude of I_{coh}, i.e. I_{corr}.

2.6.6. Single-dish spectra

For the reduction of the single-dish spectra, the data are also first masked and the background is estimated by a linear fit to the signal on the detector between a region above and below the mask so that the background "underlying" the signal can be determined.

This background subtraction is imperfect for weak sources since the linear fit is not a good description to the data. Higher order fits, on the other hand, are badly defined by the few data points that can be used for the fit and are therefore no alternative. See Figure 2.12 for an impression of the background against which the signal has to be detected.

The data reduction of the single-dish spectra has been described in detail by Tristram (2007).

2.7. Calibration and Errors

2.7.1. Calibration / Visibilities vs. correlated fluxes

The usual way of calibrating the data is by first determining the instrumental visibilities for the target and the calibrator observation $V_{\text{ins}} = I_{\text{corr}}/I_{\text{tot}}$ where I_{tot} is the total flux (in counts/s) as determined from the single-dish spectra. The calibrated visibility (amplitude) of the target is then given by

2.7. Calibration and Errors

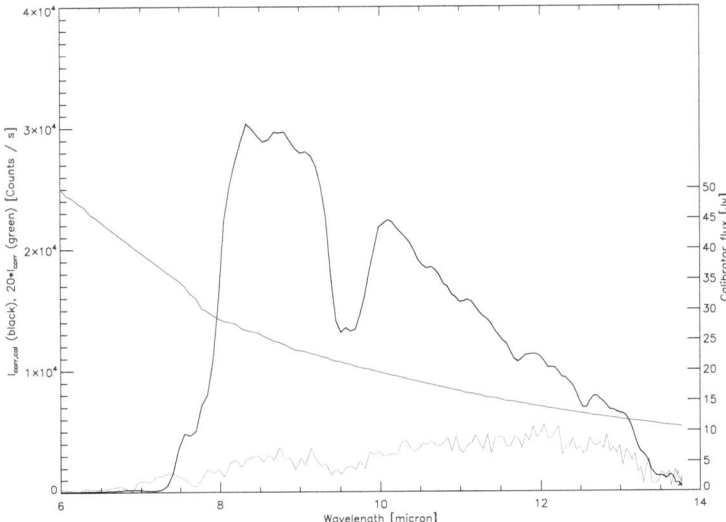

Figure 2.17.: Correlated flux count rates (in counts/s) of the calibrator $I_{\mathrm{corr,cal}}$ (black) and science target $I_{\mathrm{corr,target}}$ (green). The science target count rate has been multiplied by 20 for better readability. The spectrum of the calibrator (in Jy, see right axis) is overplotted in red. The count rate of the calibrator (black) is proportional to the product of the calibrator flux (red) and the atmospheric transfer function (Figure 2.4).

$$V = \frac{V_{\mathrm{ins,target}}}{V_{\mathrm{ins,cal}}} \cdot V_{\mathrm{cal}}, \qquad (2.34)$$

where V_{cal} is the visibility of the calibrator, usually determined from a uniform disk approximation (see subsection 2.2.6 above) to the target's intensity distribution using a diameter known from the literature (van Boekel 2004).

An alternative way is to calibrate the correlated fluxes directly without making any use of the single-dish spectra:

$$F_\nu = \frac{I_{\mathrm{corr,target}}}{I_{\mathrm{corr,cal}}} \cdot F_{\nu,\mathrm{cal}} \qquad (2.35)$$

2. Mid-infrared interferometry

where $I_{\rm corr,cal}$ is the correlated flux (in counts) of the calibrator and $F_{\nu,\rm cal}$ the flux of the calibrator (in Jy), again taken from the literature (van Boekel 2004). These quantities are shown in Figure 2.17.

Traditionally, the first method (calibrating visibilities instead of correlated fluxes) is preferred because it is less sensitive to fluctuations in the atmosphere: When calibrating visibilities, variations in the (flux) transmission of the atmosphere do not matter very much because the total flux measurements are taken usually only 5 – 10 minutes after the fringe track (correlated flux measurement). The relatively large separation between calibrator and target fringe track does not matter since, in this method, the calibrator is only used to calibrate the instrumental visibility, a quantity that depends on instrumental properties and is not expected to vary as fast as the atmosphere.

However, for weak sources such as the ones described in the following chapters, the errors of the single-dish measurements are much larger than for the correlated flux observations and it is therefore preferable to calibrate the correlated fluxes directly; otherwise the large error of a single photometry measurement (as large as 30 %!) would deteriorate the relatively well-determined correlated flux signal.

This of course requires some understanding of the atmospheric variations so that the correlated flux measurement of the calibrator can be used to reliably flux-calibrate the correlated flux of the target source.

This has been tested extensively and is described in the next subsection.

OIFITS The result and all intermediate output of EWS is stored in Optical Interferometry Flexible Image Transport System (OIFITS) format, a standardized data format for storing and exchanging visibility data (Pauls et al. 2005). Unfortunately correlated fluxes are not defined in the standard yet (as of version 1), but it is planned to include them in a second version.

2.7.2. Atmospheric stability / gains

In order to understand the atmospheric transfer function fluctuations, all calibrators from the Large Programme (see Chapter 5) were examined. These are 116 observations of 22 different stars. The standard EWS data reduction was applied and the standard EWS mask (a relatively wide mask) was shifted to the calibrator's position on the detector.

The correlated flux gain $g_{\rm corr}(\lambda)$ ("corrgain") is defined to be

$$g_{\rm corr}(\lambda) = I_{\rm corr,cal}(\lambda)/(F_{\nu,\rm cal}(\lambda) \cdot V_{\rm cal}(\lambda)), \tag{2.36}$$

where the dependency on (λ) of the above defined quantities has now explicitly been written.

The single-beam gain $g_{\rm phot}(\lambda)$ ("photgain"), on the other hand, is

$$g_{\rm phot}(\lambda) = I_{\rm tot}(\lambda)/F_{rm\nu,cal}(\lambda). \tag{2.37}$$

2.7. Calibration and Errors

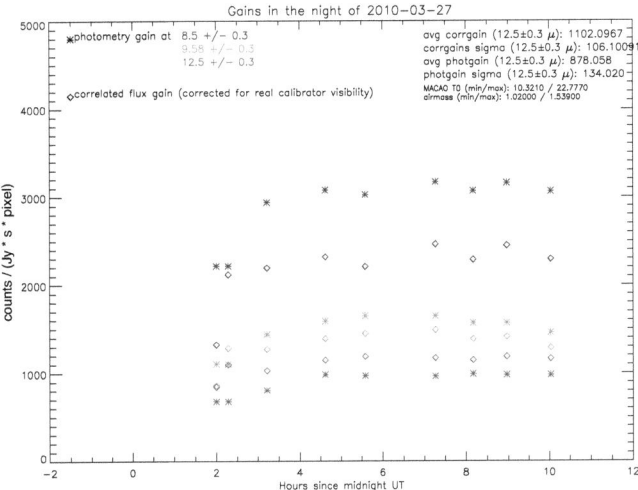

Figure 2.18.: Gains at various wavelengths for the excellent night of 2010-03-27. Some statistics about the gain variations is given in the plot.

These gains are shown for an excellent, an average and a bad night in Figures 2.18, 2.19 and 2.20, respectively. The average gains of all the Large Programme calibrators was also investigated to search for calibrator variability. They are shown in Figure 2.21.

In average and good nights, typical variations of the transfer function between a calibrator and a target observation were $\lesssim$ 5%, as determined by linear interpolation of the transfer function between the calibrator observations that surrounded the target observations. This is of the same order as the statistical error of the averaged fluxes. We tried to remove this systematic error by using these interpolated values of the transfer function to calibrate the target observations. The consistency of repeated correlated flux observations (see below) was not be increased, however, if correcting for the transmission fluctuations in this way.

Since this "gain" error is an uncertainty of the zero-point of the spectrum, this error becomes a significant contribution to the total error budget, if the statistical error (per bin) is reduced through averaging.

Atmospheric indicators for transfer function fluctuations? In bad nights, the gain can change very quickly and by a large amount. This factor limits our calibration accuracy but is so far not taken into account in the error determination.

Therefore it would be desirable to find an indicator for the gain that is measured more frequently than calibrator stars can be observed. This could potentially improve the

2. Mid-infrared interferometry

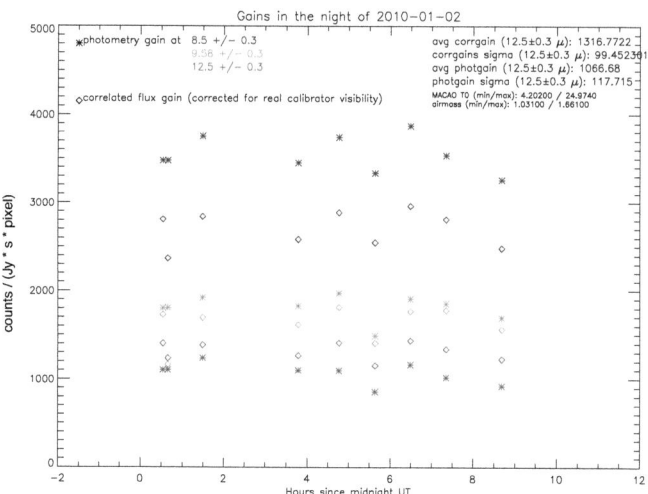

Figure 2.19.: Gains for the average night of 2010-01-02.

calibration accuracy.

As a first step, we studied the variation of corrgain with photgain, i.e. did the correlated flux transmission change in the same way as the single-dish transmission (Figure 2.22)?

2.7. Calibration and Errors

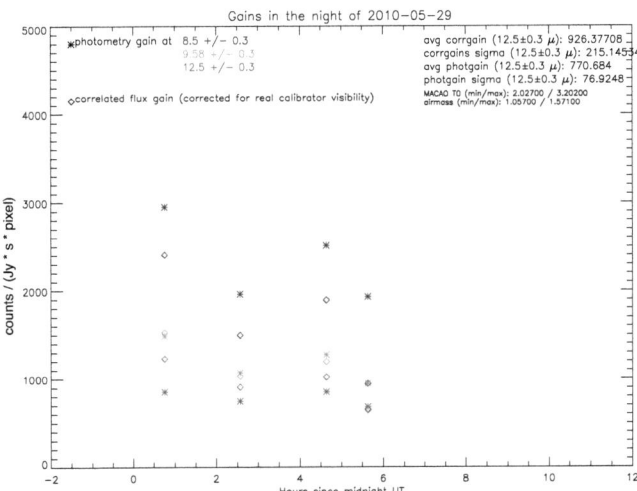

Figure 2.20.: Gains for the bad night of 2010-05-29.

2. Mid-infrared interferometry

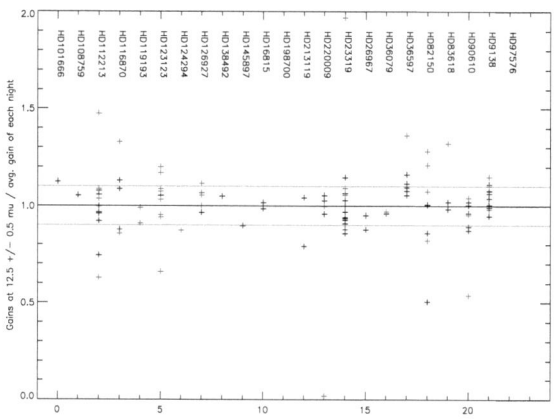

Figure 2.21.: **Test for calibrator variability** – For each calibrator measurement, the gain was calculated and divided by the average gain of the night (for this average, all calibrator measurements of the night except observations of this calibrator are taken into account). In the same way, the standard deviation (stddev) of gain per night is calculated (i.e. stddev of the gain of all calibrators in this night except the one in question). If the gain spread for the night of the calibrator measurement in question is less than 10%, it is plotted in black, otherwise in red. In this way, one would expect to have a spread of approx. 10% (horizontal lines: average + expected FWHM) in the black points, if all calibrators were good (because only those nights were selected that were stable). From this it appears that HD 112213, HD 23319 and HD 9138 are pretty much as expected, HD 36597's template might be 10% too bright, HD 90610 may be a little too faint. The extreme outlier point of HD 82150 needs a closer examination.

2.7. Calibration and Errors

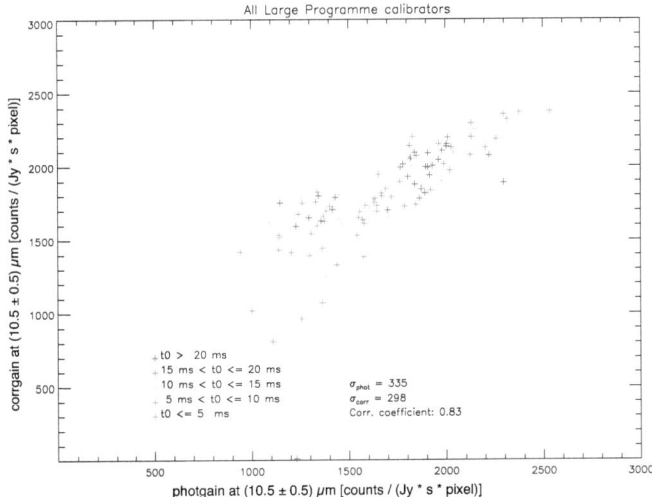

Figure 2.22.: Corr. flux gain vs. single-dish gain. The color denotes the coherence time τ_0 as determined from MACAO and statistical information about the correlation is given in the plot.

There is a good correlation (corr. coefficient = 0.83) between corrgain and photgain, meaning that, whatever causes the gain variations, affects both the single-dish and the correlated fluxes (i.e. it is not simply a correlation loss problem). This is further reinforced by looking at the correlation between coherence time and gain: The color of the data points marks the atmospheric coherence time in the visual wavelength range as seen by MACAO. Low coherence times tend to lead to somewhat lower corrgains, but there is a considerable spread.

In conclusion, no strong indicators for the atmospheric transmission of mid-infrared observations were found that are measured more frequently than the calibrator star.

2. Mid-infrared interferometry

2.7.3. Statistical error / error budget

The statistical error of the corr. flux and of the single-dish flux is estimated by splitting an observation, consisting of several thousand frames, into five equal parts and deriving the variance of these sub-observations.

For the photometries there is another source of error that is easily seen when re-observing a photometry. This error, that obviously only occurs on timescales much larger than the integration time of a single observation, is discussed in detail in Section 5.2.5.

More relevant for our studies is the question of how reliable the correlated fluxes can be determined. This cannot be as easily tested as for the single-dish measurements since the correlated flux in general can be different for every point on the (u, v) plane.

2.7.4. Systematic errors – Repeated observations

A good test for systematic errors is the re-observation of a (u, v) point on a different baseline and in a different night. This has been possible for two sources where the (u, v) plane tracks crossed each other in one or more points.

Centaurus A Due to the southern declination of Cen A, the U1U3 and U2U4 baselines cross at $(u, v) \approx (65, 50)$ m (see Figure 3.2 on page 66). This offers the possibility to observe the same (u, v) point not only on different days but also with different telescope combinations. The comparison of the correlated fluxes measured at this crossing point is displayed in Figure 2.23.

The two points that are closest together in (u, v) position (s4/s5 on U1U3, separation: 2m, yellow curves) show identical spectra. s1 on U2U4 (lower red spectrum) and the two yellow spectra are also (almost) identical within the errors – only at 8.5 micron they are slightly more separated than 1 σ. s2 on U2U4 (upper red spectrum), finally, shows a significantly larger visibility at all wavelengths than the other spectra – it is also separated by about 8m from the next nearest one. It is therefore reasonable to take this offset as real and conclude that systematic errors do not dominate over the statistical ones.

Note that between the two red spectra, a calibrator has been observed, while there has not been a re-acquisition of the source in between the two yellow curves. Since the lower red curve, taken on a different baseline and in a different night than the yellow curves, shows a very similar visibility as the yellow curves we assume that this effect is negligible.

Large Programme Among the Large Programme targets, there is one source, NGC 1365, that has been re-observed at the same (u, v) co-ordinates as before (see Figure 5.6 on page 125). The fluxes are identical within the errors – although the data have been taken three years apart (Figure 2.24)! It appears that the errors determined from the variance within one fringe track are a good estimation of the true uncertainty of the correlated flux measurements.

2.7. Calibration and Errors

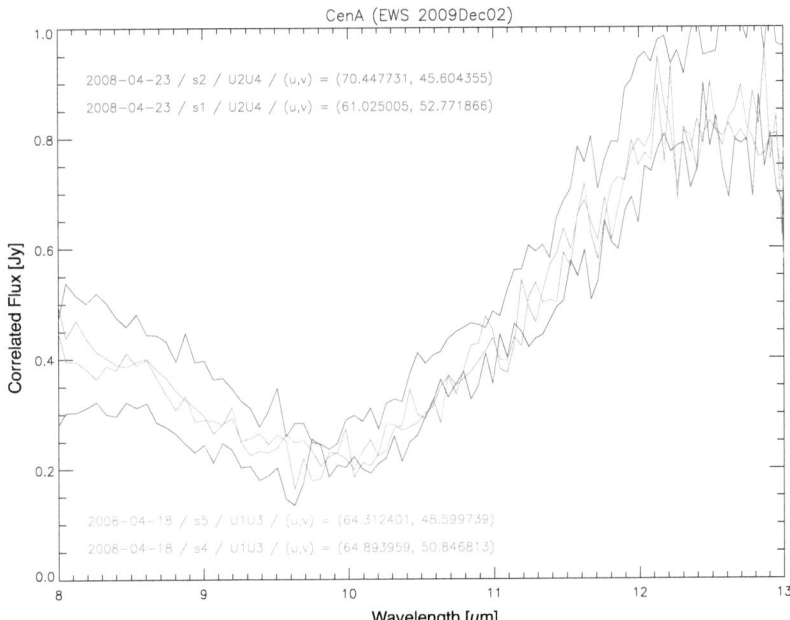

Figure 2.23.: Test for systematic errors using observations of Centaurus A (Chapter 3). The (u, v) point at which the upper red spectrum has been taken is separated by about 8m from that of the yellow spectra. The yellow spectra are separated by only about 2m in (u, v) space and the separation between either yellow spectrum and the lower red spectrum is about 5m.

2. Mid-infrared interferometry

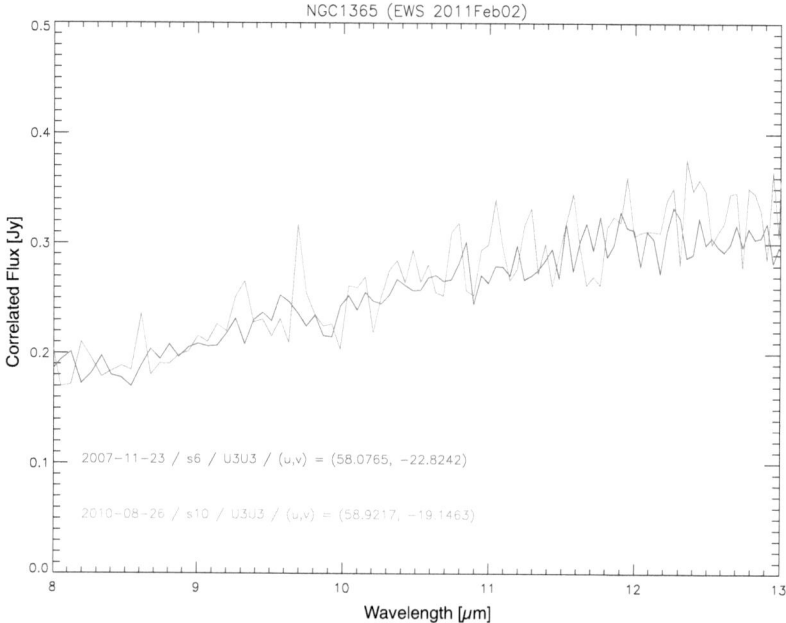

Figure 2.24.: Same as Figure 2.23, but for observations of NGC 1365 (Chapter 5).

3. Centaurus A: Dissecting the nuclear mid-infrared emission in a radio-galaxy

3.1. Introduction

3.1.1. A unique galaxy

Centaurus A (Cen A) is a unique astrophysical object that can serve as a laboratory for many sub-disciplines of (astro)physics.

While the peculiar nature of the optical "nebula" NGC 5128 was already noticed by William Herschel in 1847, only Bolton et al. (1949) identified Cen A, the largest extragalactic radio source, with the giant elliptical galaxy by using the sea interferometer at Dover Heights near Sydney, Australia.[1] Its giant radio lobes, that span about $5° \times 9°$ (270 kpc × 480 kpc) on the sky, have now been mapped in detail in large radio-synthesis images (e.g. Feain et al. 2009). Its huge size is testimony of its proximity rather than its intrinsic power: The distance to Cen A is now very well determined from a combination of various distance indicators to be just 3.8 ± 0.1 Mpc (Harris et al. 2010), making it one of the closest AGNs known. At this distance, one parsec corresponds to 54 mas. Its black hole mass of $M_{\rm BH} = 4.5^{+1.7}_{-1.0} \times 10^7 M_\odot$ (from gas kinematics, but with similar values from stellar kinematics, Neumayer et al. 2007; Cappellari et al. 2009) is rather low compared to its peers, such as M87 that has $M_{\rm BH} = (6.6 \pm 0.4) \times 10^9 M_\odot$ (Gebhardt et al. 2011).

The radio source Cen A is classified as a low luminosity radio galaxy. In the context of unified models for low-luminosity radio galaxies (see Section 1.3), Chiaberge et al. (2001) have described Cen A as a "misaligned blazar" – due to the similarity of its spectrum with that of (other) blazars – but had to employ a spine-sheath structure of the jet in order to fit the unexpectedly low values of the Doppler factor. In this scenario, the jet emission from Cen A seen by us is dominated by the slower moving sheath of the jet, but a faster moving spine (with $\gtrsim 0.45c$) was inferred from the short-time variability of individual components (Tingay et al. 1998) and actually seen from the proper motions of knots in the X-Ray jet (at ≈ 0.5 c, Hardcastle et al. 2003a). In a long-time VLBI study, Tingay et al. (2001) determined the proper motion of jet knots to be relatively slow at ≈ 0.1 c and confirmed variability of the jet components on a time-scale of months to years. This

---

[1] In a strict sense, "Cen A" therefore denotes the *radio* source and NGC 5128 the elliptical / irregular galaxy.

3. Centaurus A: Dissecting the nuclear mid-infrared emission in a radio-galaxy

variability was explained as the result of perturbations in the jet flow near the base of the jet as modeled by Agudo et al. (2001). Such a perturbation would cause enhanced synchrotron emission as it crosses slower moving shocks downstream in the jet.

Cen A is one of few AGNs that have been continually monitored for over twenty years with space-based observatories (Rothschild et al. 2011). From X-Ray observations, Evans et al. (2004) inferred an efficiency for Bondi accretion of the hot gas of $\approx 0.2\%$, lower than the expected efficiency for a standard disk (Shakura & Syunyaev 1973), but much larger than expected for Bondi accretion. Evans et al. (2004) therefore favor an accretion mode via a geometrically thin, optically thick disk for Cen A. This stands against the current view (e.g. Best 2009) that low-luminosity radio galaxies work through slow, radiatively-inefficient accretion of hot gas from their X-Ray halos. Maybe Cen A is special in this respect because the merger has accelerated the slow accretion process? Struve et al. (2010) find, though, that the kinematics of the HI disk in Cen A is very regular (down to ca. 100 pc, the resolution limit of their observations) and does not show any evidence of an accretion flow.

On the highest energy end of the spectrum, Cen A has also been detected in Very High Energy γ-rays (HESS Collaboration: F. Aharonian 2009) and due to both the extent of its radio structure and its proximity, Cen A is now discussed as one of the few plausible sites where those Ultra High Energy particles got accelerated that have been detected by the AUGER Cosmic Ray observatory (e.g. Hardcastle et al. 2009; Nemmen et al. 2010). Therefore and due to the wealth of other observations, Cen A is one of only few astronomical sources that are privileged to have become the theme of an international conference[2].

The merging history of Cen A was already evident from optical images (see Figure 3.1). The prominent dust lane is interpreted as the remainder of a disk galaxy that merged with the much larger elliptical galaxy. An impressive demonstration of the disk galaxy's impact on the elliptical galaxy was achieved by Malin et al. (1983) and Peng et al. (2002) who took very deep optical images and discovered a system of shells from which the merger was derived to have started several hundred million years ago.

The most recent comprehensive review of this source was given by Israel (1998). More recently, Morganti (2010) have summarized "the many faces of gas in Centaurus A" and Struve et al. (2010) review the timescales of the processes that might have formed Cen A's morphology. An extensive collection of Cen A resources can also be found on Helmut Steinle's webpage[3]. Images spanning a wide range of frequencies can be found at Caltech's "Cool Cosmos" webpage[4].

While some aspects of Cen A appear peculiar in comparison with other galaxies, it has already been noted by Bland et al. (1987) that these features would probably not attract our attention if Cen A were not at such a close distance to us.

---

[2] http://www.atnf.csiro.au/research/cena/
[3] http://www.mpe.mpg.de/Cen-A/
[4] http://coolcosmos.ipac.caltech.edu/cosmic_classroom/multiwavelength_astronomy/multiwavelength_museum

3.1. Introduction

Figure 3.1.: Multi-frequency composit of Cen A showing the kpc-scale jet in sub-mm (orange) and X-Rays (blue). North is up and east is left. The northern jet, which is thought to be directed towards us at an inclination of $\approx 50°$ powers the northern inner lobe; in the south the jet is not visible in the X-Rays (but in the radio, not shown here). The merging spiral and elliptical galaxies are seen in true colors (optical) and the famous cold (≈ 30 K, Weiß et al. 2008) dust lane also glows in the sub-mm regime in front of the giant elliptical galaxy. Credit: ESO/WFI (Optical); MPIfR/ESO/APEX/A.Weiss et al. (Submillimetre); NASA/CXC/CfA/R.Kraft et al. (X-ray)

3.1.2. Cen A in the infrared

The nucleus of Cen A is attenuated by $A_V \gtrsim 8$ mag as determined from an SED fit to HST photometry of the nucleus by Marconi et al. (2000); these authors also marginally detected the core in the V band. In order to study the nucleus of Cen A in the optical regime in detail, however, one must observe in the near- or mid-infrared.

Let us start with an overview of the infrared morphology of Cen A, zooming in from the kpc-scale. There, Spitzer observations have been successfully modeled as a warped disk ($r \approx 100''$, 2 kpc) of gas and dust (Quillen et al. 2006), following the original model by Bland et al. (1987). The disk is seen nearly edge-on and warps crossing our line of sight are thought to be responsible for bright absorption (near-IR) or emission (mid-IR) knots. There appears to be a gap in this disk at $6'' \lesssim r \lesssim 50''$, corresponding to ≈ 0.1 kpc $\lesssim r \lesssim$ 0.8 kpc. Zooming further in, a 500 pc dust shell is seen in a Spitzer spectrographic map (Quillen et al. 2008). In the vicinity of this shell, the molecular hydrogen is found to have temperatures of 250 – 720 K, higher than in non-active galaxies, but similar to values found in Seyfert galaxies. At this distance, highly ionized species are found at position angles near the jet that are probably excited by radiation from the nucleus.

Neumayer et al. (2007) used adaptive optics assisted integral field spectroscopy in the near infrared to derive a model for the mass distribution of the central $3''$ (60 pc) from the observation of molecular hydrogen emission. At their resolution limit of 120 mas (ca. 2 pc) in the K band, they find a PA of $(148.5 \pm 1)°$ and an inclination of $(37.5 \pm 2)°$ for the velocity field in their tilted ring model.

Espada et al. (2009) observed the central 1 arcmin (1 kpc) of Cen A in the ^{12}CO$(2-1)$ emission line with the Submillimeter Array with a resolution of 100 pc × 40 pc. Using a warped thin disk model, they find that this molecular gas emission in the inner $20''$ (400 pc) is elongated at a PA of $\approx 155°$ and they interpret the observed axis-ratio of the presumably circular disk as an inclination of $i = 70°$.

Attempts to resolve the core in the infrared have been made by Schreier et al. (1998) who found an unresolved (< 100 mas) nucleus in the K band using HST WFPC data. Using observations with HST NICMOS, Capetti et al. (2000) found a 3 σ upper limit on the core FWHM of 0.8 pc and found a high polarization of $(11.1 \pm 0.2)\%$, difficult to explain with dust scattering as it would require a much larger column density of dust (corresponding to $A_V \approx 50$) than is observed.

High-resolution ground-based observations of the nucleus of Cen A using the 8-10 m class observatories, finally, still find an unresolved nucleus in the mid-IR with upper limits of 190 mas at 8.8 micron (Radomski et al. 2008) or detect it as a point source (Horst et al. 2009; van der Wolk et al. 2010)

Taken together, long-baseline optical interferometry is obviously needed for getting closer to the infrared core of Cen A.

Fortunately, its relatively bright mid-IR core of $F_{\text{core}}(12\mu\text{m}) \approx 1.5$ Jy (variable) puts it in the range of the MIDI sensitivity. Cen A is the only radio galaxy that can be studied with MIDI in great detail.

With limited (u, v) coverage, Meisenheimer et al. (2007) found from earlier MIDI data that the mid-IR emission from the central parsec of Cen A is dominated (80% at 8 μm and 60% at 13 μm, respectively) by an unresolved point source (diameter < 10 mas). By comparison with multi-wavelength data, this emission was classified as being most likely non-thermal (synchrotron) in origin. The size of the resolved emission was estimated to be $\gtrsim$ 30 mas (0.6 pc) roughly perpendicular to the axis of the radio jet and $\lesssim$ 12 mas (0.2 pc) along the axis of the radio jet, respectively. The position angle of the major axis was approximated to be about 127 $\pm$ 9°. It was found to be consistent with a geometrically thin, inclined dusty disk. With only 4 (u, v) points, the PA of the inferred disk was fixed to the direction orthogonal to the PA of the maximum visibilities.

Here we report results from a more extensive set of (u, v) points that allows to fit model source brightness distributions to the nuclear mid-infrared emission and determine the parameters of the possible emission components more precisely.

3.2. Instrument, observations and data reduction

Observations were performed in the N band (8μm $< \lambda <$ 13μm) with MIDI using the observing procedure described in Chapter 2.

For the observation log as well as the observational parameters, see Table 3.1. In total, 34 fringe track observations were made between 2004 and 2010. 13 of these observations are unusable due to various reasons. This leaves a total of 21 successful fringe track observations in two epochs: Four in 2005 and 17 in 2008. Of the ones that were observed in 2008, 7 are practically duplicates in the (u, v) plane and were taken either to increase signal-to-noise or to check for systematic errors.

These observations provide effective spatial resolutions $\lambda/(3BL)$ (see Section 2.2.5) of $\approx 5 - 11$ mas at 8.5 μm and $\approx 7 - 16$ mas at 12.5 μm, respectively. Calibrators were selected to be close in airmass. Due to the orientation of the VLTI baselines and the declination of the source, there are unfortunately no long UT baselines in the lower left (and upper right) quadrant of the (u, v) plane (see Figure 3.2).

The data reduction procedure was as described in Section 2.6[5].

The 2005 data were re-reduced with a more recent EWS version (in comparison to the one used by Meisenheimer et al. (2007)). The results differ slightly from the previously published data reduction but the relative flux levels – between the various (u, v) points – remain the same.[6]

The 2005 data were first published in Meisenheimer et al. (2007). Averaged 12.5 μm visibilities of the 2008 data were reported in Burtscher et al. (2010).

Unsuccessful observations of NGC 5128 For completeness, the unsuccessful observations and observation attempts are listed below. Six fringe tracks had to be discarded

---
[5]For the data reduction of this source, the nightly build from Dec 02 2009 was used.

[6]The data reduction was also compared to a data reduction with an even more recent EWS version (2010Nov20, including the so called "weak sources routines", see Section 2.6). Within the errors, the results were identical.

Table 3.1.: Overview of all successful MIDI Cen A observations and related calibrators. Observation id, projected baseline length BL and position angle PA, effective spatial resolution of the interferometer $\Theta_{min} = 8.5\mu m/3BL$, Airmass and Seeing

Date and Time [UTC]		id	BL [m]	PA [°]	Θ_{min} [mas (pc)]	Airmass	Seeing[a] ["]	Associated calibrator
2005-02-28:	UT3 – UT4							
06:07:16		s4	58.22	96.41	10.0 (0.19)	1.11	0.54	HD112213 (c7)
08:29:03		s6	62.37	119.79	9.4 (0.18)	1.08	0.87	HD112213 (c9)
2005-05-26:	UT2 – UT3							
00:16:26		s3	46.51	27.96	12.6 (0.24)	1.12	0.68	HD112213 (c3)
02:27:21		s6	44.11	46.72	13.3 (0.25)	1.07	0.69	HD112213 (c5)
2008-04-18:	UT1 – UT3							
02:54:31		s1	101.35	22.43	5.8 (0.11)	1.10	1.07	HD112213 (c6)
04:51:38		s2	96.72	38.11	6.1 (0.11)	1.06	1.05	HD112213 (c6)
05:03:06		s3	95.92	39.48	6.1 (0.12)	1.07	1.10	HD112213 (c6)
07:03:41		s4	82.44	51.92	7.1 (0.13)	1.28	1.71	HD119193 (c8)
07:15:19		s5	80.61	52.92	7.3 (0.14)	1.32	1.22	HD119193 (c8)
2008-04-19:	UT1 – UT4							
02:09:13		s2	128.74	40.33	4.6 (0.08)	1.17	0.95	HD112213 (c3)
02:20:33		s3	129.17	42.48	4.6 (0.08)	1.15	1.10	HD112213 (c3)
2008-04-21:	UT3 – UT4							
01:47:30		s1	53.46	87.27	10.9 (0.20)	1.20	0.66	HD111915 (c2)
03:55:10		s4	61.85	108.28	9.5 (0.18)	1.06	0.65	HD112213 (c4)
05:26:40		s5	62.02	123.96	9.5 (0.18)	1.10	0.60	HD112213 (c4)
05:38:02		s6	61.77	126.06	9.5 (0.18)	1.12	0.68	HD112213 (c4)
08:48:03		s9	55.44	169.34	10.5 (0.20)	1.98	0.57	HD119193 (c7)
08:59:20		s10	55.27	172.34	10.5 (0.20)	2.12	0.54	HD119193 (c7)
2008-04-23:	UT2 – UT4							
01:06:08		s1	80.68	49.15	7.3 (0.14)	1.30	1.17	HD119193 (c4)
01:42:35		s2	83.92	57.08	6.9 (0.13)	1.20	0.69	HD111915 (c5)
04:52:34		s8	87.18	91.22	6.7 (0.12)	1.08	1.00	HD112213 (c7)
06:16:51		s9	78.88	106.62	7.4 (0.14)	1.21	1.08	HD119193 (c10)

[a] at 0.5 μm from the VLT seeing monitor

3.2. Instrument, observations and data reduction

because the telescopes were still in chopping state. This data is hard to reduce because the chopping state is not stored in the fringe track files (in MIDI's HIGH SENSE mode), making it hard to flag the "sky" frames. Apart from that, even if the data could be reduced, the reduced data would be unreliable as it is unclear what effect the chopping has on the estimation of the correlated flux. Chopping during HIGH SENSE mode is not intended, but it apparently occurs sometimes when an observation template is not completely executed (e.g. if "Phot B" is aborted). In future observations one must take care to verify the chopping state before starting the fringe track observation.

- 2004-06-01 (43m, 43°, UT2 – UT3), 02:31:28: no fringe found
- 2005-05-26 (46m, 31°, UT1 – UT3), 00:37:35 (s4): M2 was chopping during fringe track
- 2006-01-20 (130m, 49°, UT1 – UT4), 08:50:46 (s1): fringe was not tracked
- 2008-04-16 (99m, 33°, UT1 – UT3), 04:05:48 (s1): very bad seeing, very low signal
- 2008-04-18 (54m, 61°, UT1 – UT3), 09:26:26 (s6): very high airmass, very low signal
- 2008-04-19 (114m, 79°, UT1 – UT4), 06:08:06 (s8): M2 was chopping during fringe track, photometry useable
- 2008-04-19 (113m, 80°, UT1 – UT4), 06:11:55 (s9): M2 was chopping during fringe track, photometry useable
- 2010-05-30 (72m, 117°, UT2 – UT4), 04:39:38 (s3): experimental GRISM observation
- 2010-05-30 (71m, 118°, UT2 – UT4), 04:47:32 (s4): M2 was chopping during fringe track
- 2010-05-30 (70m, 121°, UT2 – UT4), 04:57:57 (s5): M2 was chopping during fringe track
- 2010-05-30 (69m, 122°, UT2 – UT4), 05:01:40 (s6): M2 was chopping during fringe track
- 2010-05-30 (62m, 134°, UT2 – UT4), 05:45:59 (s7): bad track
- 2010-05-30 (61m, 136°, UT2 – UT4), 05:52:22 (s8): bad track

3.3. Results

3.3.1. (u, v) coverage

The resulting (u, v) coverage is displayed in Figure 3.2. It was conceived to be as uniform as possible. Even when using all available UT baselines, however, the south-eastern (= north-western) quadrant of the (u, v) coverage is less uniformly sampled than the north-eastern (= south-western) quadrant due to the orientation of the VLT Unit Telescopes. Also, some of the planned observations failed for technical reasons.

We observed visibilities on virtually identical (u, v) points and in the same epoch (2008) at the crossing point between the U1U3 and U2U4 baselines and re-observed, on the U3U4 baseline, some of the visibilities observed in 2005. This observing strategy proved to be a very important test for systematics and variability.

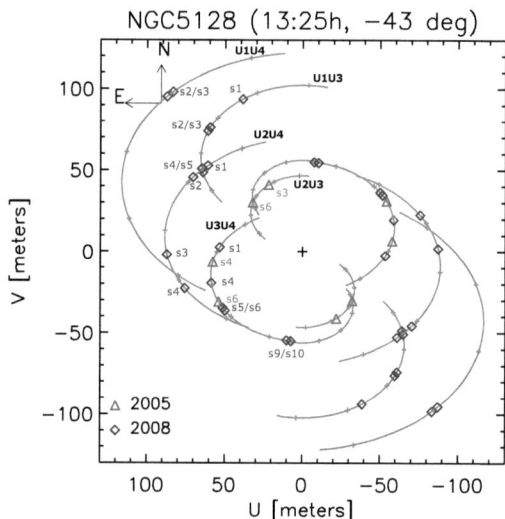

Figure 3.2.: (u, v) coverage of the successful 2005 and 2008 MIDI observations of Cen A. Every visibility appears twice in this plot (because the telescope positions are interchangeable). The solid lines are the (u, v) positions traced by the various baselines (UT combinations are labelled) as a result of earth's rotation. (u, v) plane tracks are followed counter-clockwise (CCW) if $\delta < 0$, clockwise (CW) if $\delta > 0$. They are truncated at telescope elevations of 30 degrees above the horizon, the elevation limit of the VLT Unit Telescopes. Small grey crosses denote Hour Angles = -4, -2, 0, 2, 4.

3.3.2. Correlated fluxes

The resulting calibrated correlated fluxes are displayed in the left panels of Figure 3.3. Besides the deep silicate absorption feature, the most prominent result of the correlated fluxes is that they are always much larger than our detection limit ($\approx$ 50 mJy). At 8.5 μm, for example, they are $\gtrsim$ 300 mJy, even on the longest baselines.

3.3.3. Single-dish spectra

Single-dish spectra were taken after every correlated flux observation. The accuracy of these observations is limited by imperfect thermal background reduction and they have a much larger scatter than the correlated flux measurements. We show both the 2005 and 2008 single-dish spectra in Figure 3.4. The 2005 spectra differ (within the errors) from the ones shown in Meisenheimer et al. (2007) since they have been reduced using a slightly different data reduction technique (especially: a different mask), as described in Section 2.6.

All Cen A single-dish spectra (per epoch, i.e. for 2005-02, 2005-05 and for 2008-04) were stacked and combined to one average spectrum. This is possible since only the correlated flux depends on the projected baseline length and position angle and it is not expected that the N band source changes within a few days (the 2008 observations were scheduled within one week to exclude variability). The single-dish flux of 2005-05-26 was found to be too low because of background subtraction problems (see Figure 3.4). It was discarded for the further analysis.

The single-dish spectra are dominated by a broad silicate absorption feature. They also show emission from the [Ne II] 12.81 μm forbidden line, that is absent in all correlated flux spectra and most likely originates from the dust shell discovered by Quillen et al. (2008).

3.3.4. Visibilities

Finally, the calibrated correlated spectra were divided by the calibrated and averaged single-dish spectra to derive calibrated visibilities. In contrast to the usual technique to determine the visibility for each measurement by dividing the correlated flux by the subsequently taken single-dish spectrum, we use our averaged single-dish spectra (of the relevant epoch) to derive the visibilities. This greatly reduces the statistical error of the visibility at the cost of an increased systematic calibration uncertainty. Since the nights of observation were stable in terms of total mid-infrared atmospheric throughput, the additional systematic error introduced by averaging total fluxes over several nights is considered to be small compared to the gain in statistical precision (see Section 2.7 for further discussion of calibration issues). These visibilities are shown in the right panels of Figure 3.3. For the two observations on 2005-05-26, the single-dish fluxes observed on 2005-02-28 were taken to derive visibilities, since the 2005-05-26 single-dish measurements are not correct (see Figure 3.4).

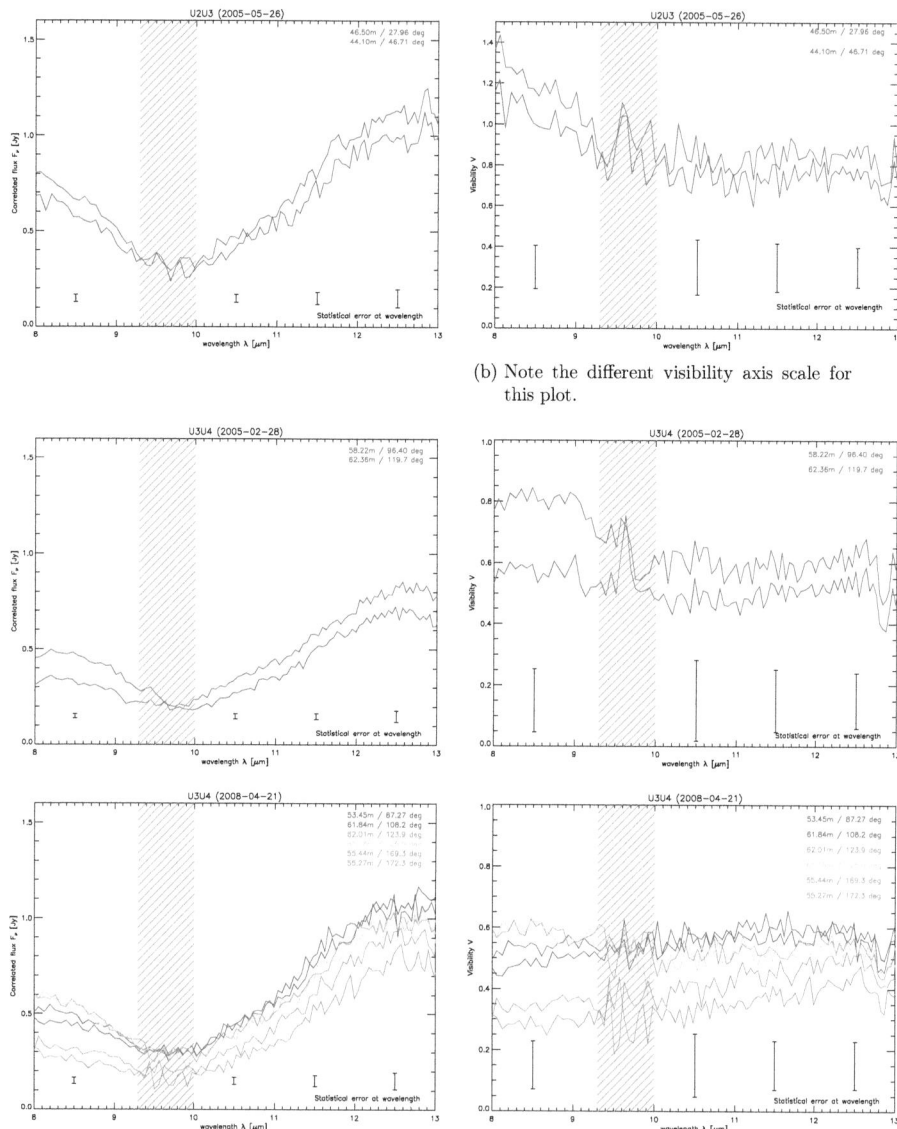

Figure 3.3.: MIDI observations of Cen A of 2005 and 2008. The days of night begin and baselines are given in the plot headers. Left: Correlated Fluxes, Right: Visibilities (using averaged photometries), see text for details. Note that the U3U4 baseline has been used for observations in both 2005 and 2008. The spectra taken on this baseline in 2005 and in 2008 are shown next to each other for better comparability. The average errors of all spectra shown in each graph are displayed at the bottom of each plot at four wavelengths (the error bars are $\pm 1\sigma$). The region of the telluric Ozone absorption feature is hatched (see Section 2.3). *Continued on next page*

3.3. Results

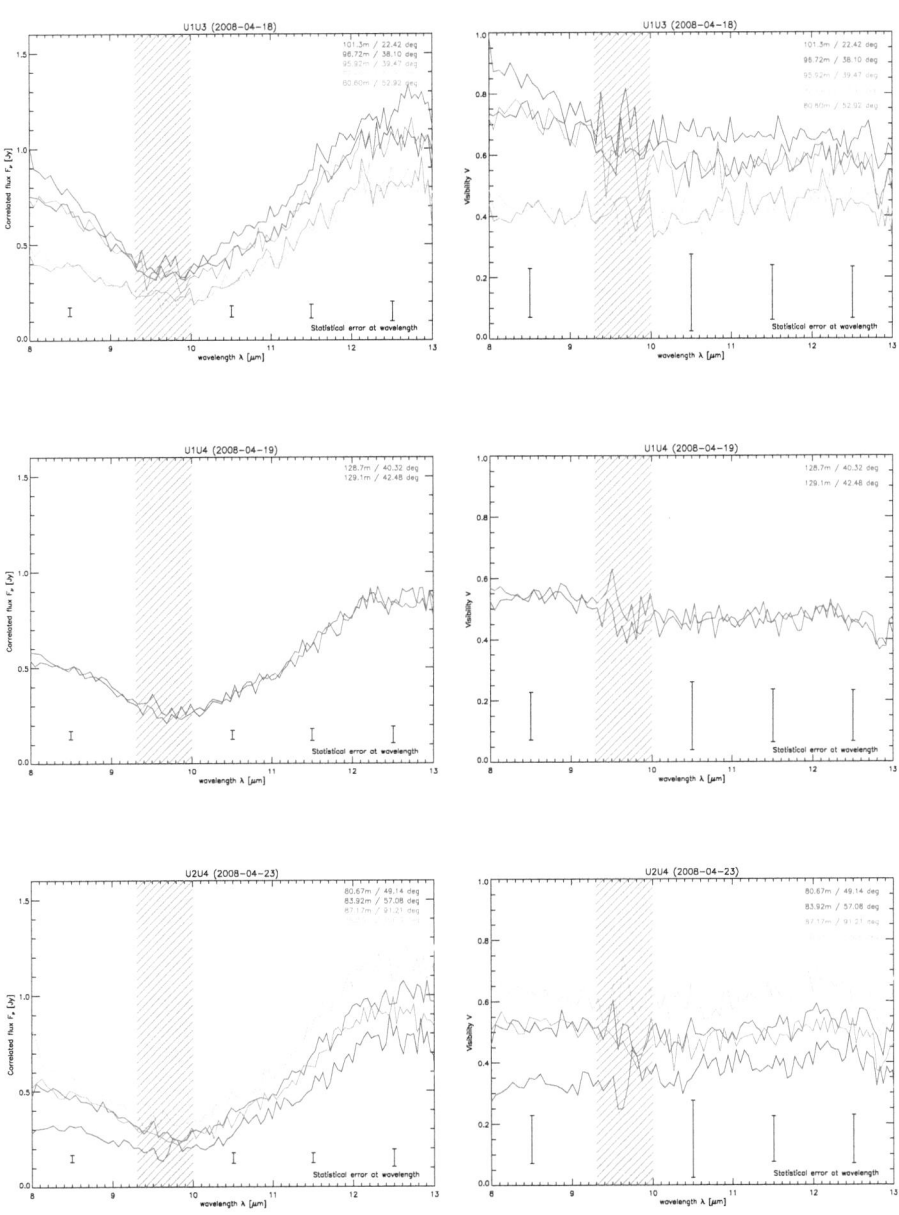

Figure 3.3.: — *Continued*

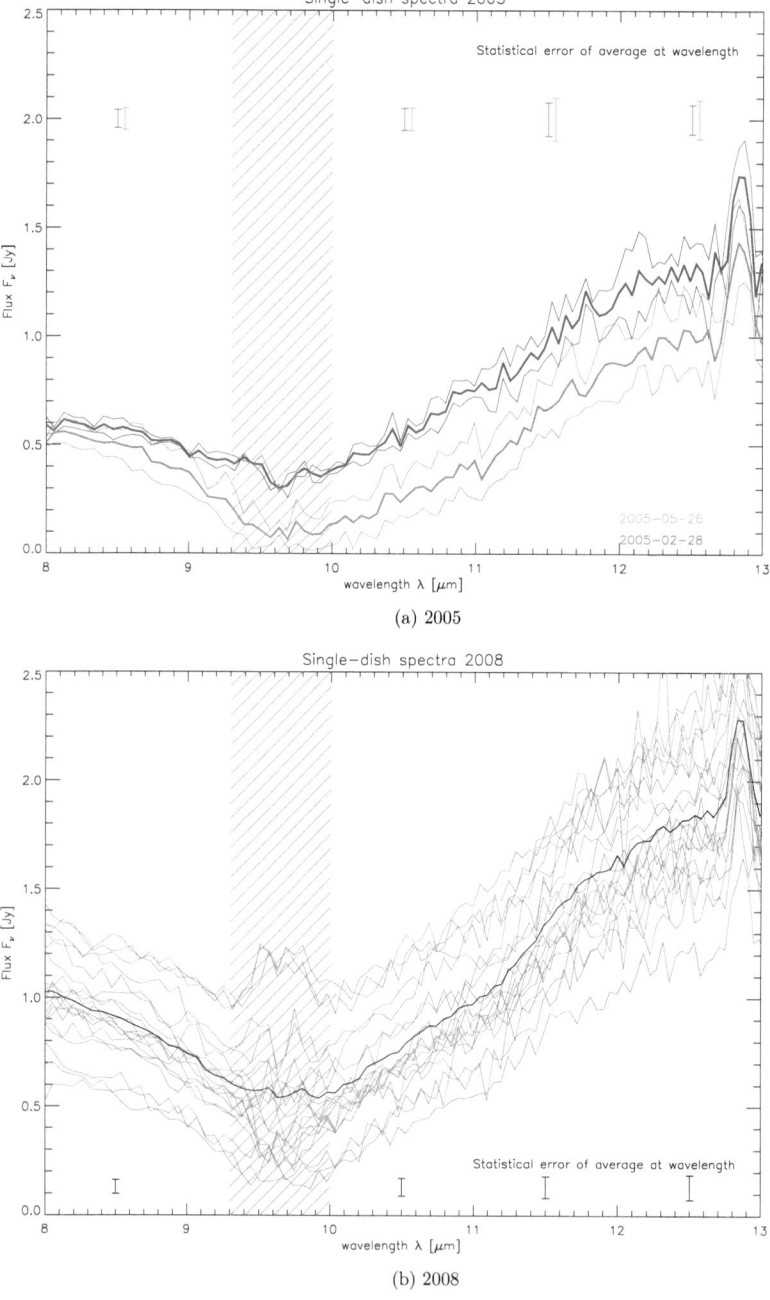

Figure 3.4.: Single-dish (total flux) observations of Cen A in 2005 (a) and in 2008 (b). The dates of observations in 2005 are given in the plot, in 2008 the individual observations are shown in grey and the averaged total flux in black. The statistical errors at selected wavelengths are plotted on top and below the spectra for each epoch (the error bars are $\pm 1\sigma$). The single-dish spectra of 2005-05-26 were discarded due to their low flux and incorrect shape.

3.3. Results

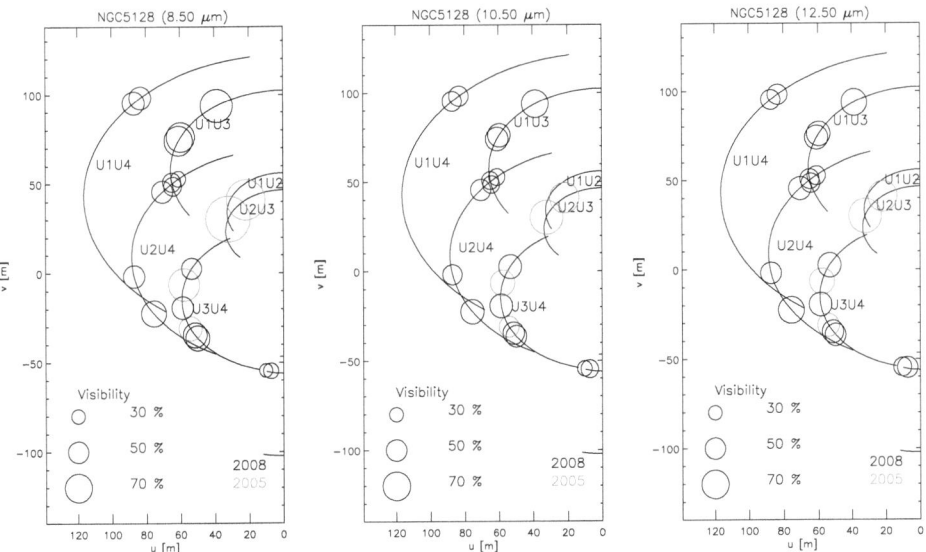

Figure 3.5.: Visibilities from the 2005 and 2008 epochs in the (u,v) plane at 8.5 ± 0.2 μm, 10.5 ± 0.2 μm and 12.5 ± 0.2 μm. Each symbol represents a visibility measurement, the radius denotes the visibility amplitude. Note that only half of the (u,v) plane is shown (the other half is point-symmetric with respect to the origin to the one shown).

For this work, the visibilities were then averaged at 8.5 ± 0.2 μm, 10.5 ± 0.2 μm and 12.5 ± 0.2 μm. They are given in Table 3.2 and visualized in the (u,v) plane in Figure 3.5 where the radius of each ring represents the visibility amplitude of the observation at these (u,v) coordinates.

The error of the calibrated, averaged visibility is composed of the statistical error of the correlated flux and that of the averaged photometry (see Section 2.7) as well as the error of the calibration templates (assumed to be 5 %, R. van Boekel, priv. comm.). These errors were propagated using the rules of propagation of uncertainty (e.g. Barlow 1989, Chap. 4.3).

Systematic errors were studied extensively and also specifically for this source. The investigation is explained in Section 2.7 and the main result is, from re-observations of a point on the (u,v) plane, that systematic errors do not dominate over statistical errors for these observations.

71

Table 3.2.: Averaged visibilities of the 2005 and 2008 observations of Cen A, used for the model fits described below.

id	u [m]	v [m]	BL [m]	PA [°]	$V(\lambda_0 \pm 0.2\mu m)$ $\lambda_0 = 8.5\ \mu m$	$\lambda_0 = 10.5\ \mu m$	$\lambda_0 = 12.5\ \mu m$
\multicolumn{8}{c}{2005-02-28 U3U4}							
s4	57.9	-6.5	58.2	96.4	0.811 ± 0.035	0.613 ± 0.043	0.627 ± 0.030
s6	54.1	-31.0	62.4	119.8	0.575 ± 0.039	0.483 ± 0.048	0.539 ± 0.032
\multicolumn{8}{c}{2005-05-26 U2U3}							
s3	21.8	41.1	46.5	28.0	1.188 ± 0.055	1.550 ± 0.096	0.994 ± 0.045
s6	32.1	30.2	44.1	46.7	1.371 ± 0.050	1.749 ± 0.091	1.130 ± 0.040
\multicolumn{8}{c}{2008-04-18 U1U3}							
s1	38.7	93.7	101.3	22.4	0.827 ± 0.025	0.679 ± 0.035	0.657 ± 0.022
s2	59.7	76.1	96.7	38.1	0.728 ± 0.028	0.556 ± 0.036	0.583 ± 0.026
s3	61.0	74.0	95.9	39.5	0.732 ± 0.027	0.560 ± 0.041	0.570 ± 0.025
s4	64.9	50.8	82.4	51.9	0.433 ± 0.036	0.396 ± 0.046	0.454 ± 0.032
s5	64.3	48.6	80.6	52.9	0.424 ± 0.044	0.389 ± 0.052	0.441 ± 0.033
\multicolumn{8}{c}{2008-04-19 U1U4}							
s2	83.3	98.1	128.7	40.3	0.540 ± 0.026	0.469 ± 0.039	0.470 ± 0.025
s3	87.2	95.3	129.2	42.5	0.547 ± 0.029	0.458 ± 0.038	0.459 ± 0.025
\multicolumn{8}{c}{2008-04-21 U3U4}							
s1	53.4	2.5	53.5	87.3	0.487 ± 0.034	0.566 ± 0.033	0.584 ± 0.022
s4	58.7	-19.4	61.8	108.3	0.534 ± 0.028	0.564 ± 0.033	0.561 ± 0.023
s5	51.4	-34.6	62.0	124.0	0.595 ± 0.025	0.524 ± 0.035	0.537 ± 0.024
s6	49.9	-36.4	61.8	126.1	0.599 ± 0.027	0.506 ± 0.037	0.526 ± 0.023
s9	10.3	-54.5	55.4	169.3	0.279 ± 0.041	0.329 ± 0.048	0.424 ± 0.028
s10	7.4	-54.8	55.3	172.3	0.342 ± 0.036	0.410 ± 0.038	0.493 ± 0.025
\multicolumn{8}{c}{2008-04-23 U2U4}							
s1	61.0	52.8	80.7	49.1	0.336 ± 0.041	0.369 ± 0.045	0.426 ± 0.028
s2	70.4	45.6	83.9	57.1	0.515 ± 0.027	0.494 ± 0.037	0.546 ± 0.022
s8	87.2	-1.8	87.2	91.2	0.527 ± 0.031	0.462 ± 0.038	0.500 ± 0.024
s9	75.6	-22.6	78.9	106.6	0.628 ± 0.029	0.585 ± 0.038	0.644 ± 0.023

3.3. Results

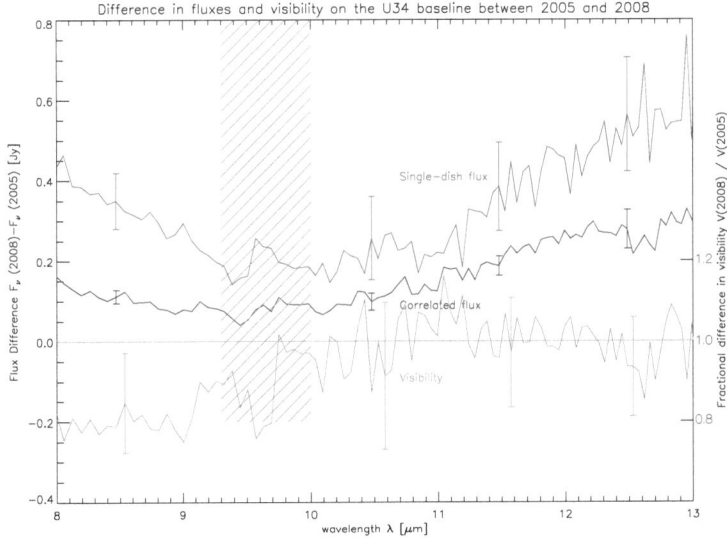

Figure 3.6.: Difference of the correlated (blue curve) and single-dish fluxes (red curve) and visibilities (green curve) between the 2005 and 2008 observations. For the correlated fluxes and visibilities, the average of the fringe tracks s1, s4, s5, s6 for the observations on 2008-04-21 and the average of the fringe tracks s4 and s6 (on 2005-02-28) were taken into account. They were observed at very similar (u, v) points (see Figures 3.2 and 3.5). The difference in visibility is consistent with no change (green straight line). The region of telluric ozone absorption is hatched as it is untrustworthy, especially in the single-dish fluxes and the visibilities. The error bars of the three curves are slightly offset for clarity.

3.3.5. Variability

Cen A clearly increased its mid-infrared flux between our 2005 and 2008 observations. This is most directly seen by comparing the total fluxes between the two epochs (Figure 3.4). The source increased its total flux by $\approx$ 350 mJy at 8.5 μm, $\approx$ 250 mJy at 10.5 μm and $\approx$ 550 mJy at 12.5 μm.

The correlated flux also increased. However, comparing correlated fluxes requires some caution. Since they, in general, depend on (u, v) coordinates, one must take observations at sufficiently similar (u, v) coordinates to compare them over time. More precisely: The maximum separation two (u, v) points may have to still show identical correlated fluxes

3. Centaurus A: Dissecting the nuclear mid-infrared emission in a radio-galaxy

depends on the structure of the source. For an unresolved or over-resolved source, the correlated flux is identical at all (u, v) points (namely V=1 or V=0, respectively); for a complex source, though, it may vary already between two points that are separated by about one telescope diameter. Since the detailed structure of the source was unknown in advance, we re-observed three (u, v) points on the U3U4 baseline in 2008 (s1, s4, and s5/s6) that are interlaced with two observations that were taken in 2005 (s4 and s6, see Figure 3.2). As can be seen from Figure 3.3, the two red spectra of the 2005 U3U4 measurements are comparable to each other and the same is true for the three orange/red spectra of the 2008 U3U4 measurements, i.e. no changes in the visibility occur for small separations in the (u, v) plane. In order to compare them, we therefore compute from them an averaged 2005 and an averaged 2008 correlated flux of this (u, v) region and plot the difference. This is shown in Figure 3.6. The increase of correlated flux at ≈ 60m/$100°$ in April 2008 with respect to February / May 2005 is ≈ 100 mJy at 8.5 μm and 10.5 μm and ≈ 250 mJy at 12.5 μm.

In terms of visibility change, the source's brightness increase is consistent with no change in visibility at all wavelengths, i.e. the correlated flux increased roughly by the same *factor* as the total flux.

3.4. Modelling

3.4.1. Considerations for model fitting

With phase-less data on sparsely sampled (u, v) planes (see Figure 3.2), it is impossible to reconstruct meaningful images directly, i.e. images that show more than just the properties of the synthesized beam. We therefore model the source brightness distribution (the "image") with simple geometrical components (e.g. point sources, Gaussians, rings, ellipses etc.) and constrain the parameters of the model with the observed visibilities, see Section 2.2.6.

Here we present our model fit to the averaged visibilities at 8.5, 10.5 and 12.5 μm. The wavelengths were chosen to be safely outside the low signal/noise regions at the edges of the N band and outside of the region of ozone absorption. They also avoid the [Ne II] 12.81 μm forbidden line that is seen in the single-dish spectra. The fits were performed with Lyon's Interferometric Tool prototype (LITpro), an interferometric model fitting tool, provided by the Jean-Marie Mariotti Center (JMMC, Tallon-Bosc et al. 2008)[7]. It is based on a modified Levenberg-Marquardt algorithm to find the least χ^2 value (see Section 2.2.7) within given parameter boundaries and provides a graphical user interface to examine the (u, v) plane with a number of geometrical models.

Since the brightness increase between the 2005 and 2008 observations could have caused a change in the structure of the source, we treat the two epochs as probing separate source states and only take into account the data from the 2008 epoch for the geometric fits. We cannot perform the same kind of fit for the 2005 observations since this exploratory

---
[7]The online version of LITpro can be found at http://www.jmmc.fr/litpro_page.htm

observation sampled the (u,v) plane too sparsely to constrain model parameters in a geometrical model fit.

From examining the visibilities of the 2008 epoch on the (u,v) plane (Figure 3.5), it is not obvious which structure they probe. To nonetheless get a reliable impression of the source brightness distribution responsible for the visibility pattern, we will approach it with a set of geometrical models based on the following observations:

- The visibilities are not very different over the entire (u,v) plane. They are significantly larger than 0 and lower than 1 at every point. This suggests a combination of an (unresolved) point source and an (over-resolved) larger source.

- The visibilities in the north-east quadrant ($0° < $ PA $ < 90°$) tend to be larger than the ones in the south-east quadrant, at $90° < $ PA $ < 180°$. This effect is strongest at 8.5 μm and could indicate an elongated structure.

- Furthermore, and again most readily seen at 8.5 μm, not all visibilities in the north-east are larger than the ones in the south-east. Rather, they seem to show a co-sinusoidal pattern in the (u,v) plane, centered at $(0,0)$, with a period of roughly 50m. Its normal is at a position angle of ca. 45° (see below for a more detailed explanation and figures).

- Between 2005 and 2008 the source brightened considerably, but the visibility (on one baseline) stayed more or less constant (see Figures 3.5 and 3.6). We will discuss potential scenarios for the luminosity increase in Section 3.5.

3.4.2. Geometrical models for the surface brightness distribution

Starting from these observations, we will add complexity to the simple most model by adding (free) parameters until we reach a satisfactory fit with χ_r^2 of the order of 1.

The model geometries are shown in Figure 3.7 and the coordinate system used to describe the locations of the components in real space is (x, y) = (RA, DEC) in units of mas. Analytic transforms of some of the "building block" functions which constitute the models are given in Section 2.2.6. A critical discussion of the model fits is given in Section 3.5.

Model a: Point source + concentric circular Gaussian The most simple explanation of the visibilities, model a, consists of a point source that contributes the flux fraction f_p (flux F_p) and a circularly symmetric Gaussian with a FWHM of Θ_g that contributes the flux fraction f_g (flux F_g) to the total flux (see Figure 3.7 for a sketch of the model). The two sources are concentric, i.e. Δx, Δy are fixed at 0.

In the (u,v) plane, this transforms to a Gaussian drop-off of the visibilities from 100 % at $(u,v) = (0,0)$, asymptotically approaching f_p at baselines $\gg \lambda/\Theta_g$.

The (u,v) plane fit is visualized in Figure 3.8 and the best-fitting parameters for this model are given in Table 3.3.

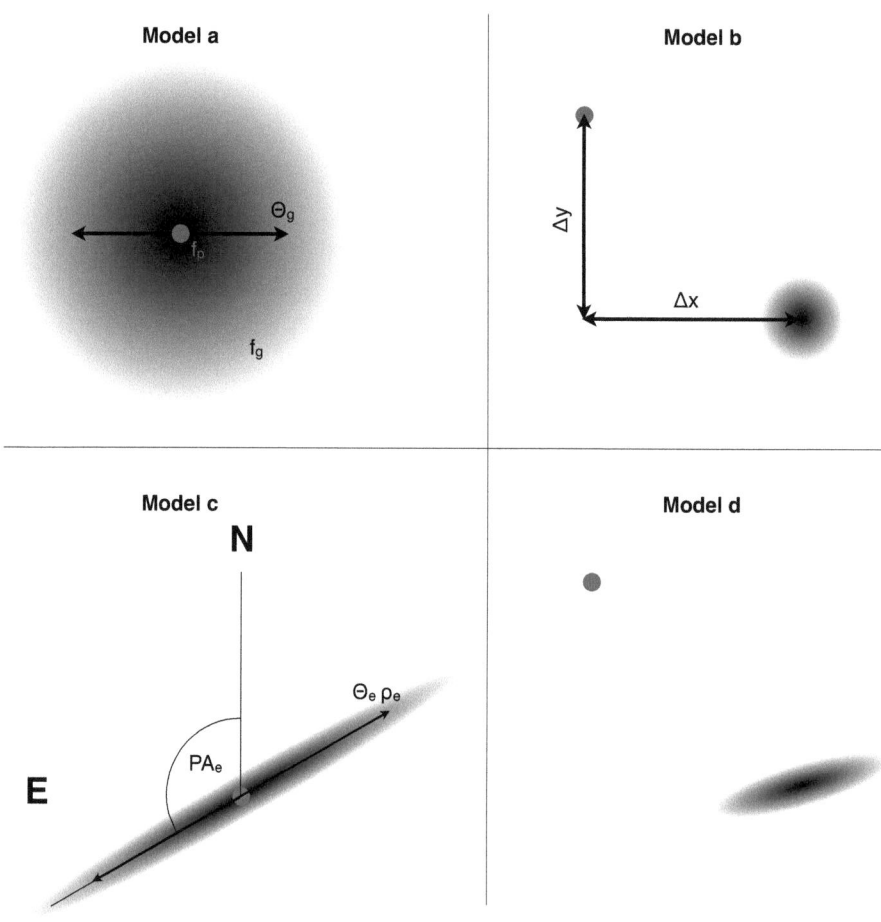

Figure 3.7.: Geometrical models for the visibilities (here in real space). The fitted parameters are marked in the model images (where they are introduced) and explained in Section 3.4 for the various models.

Table 3.3.: Best-fitting parameters for the fits of the geometrical models a and b to the averaged visibilities of Cen A at 8.5, 10.5 and 12.5 μm. Fixed parameter values are given in brackets. For convenience, the flux contributions of the two components are also given in units of Jy (F_p and F_g). See the text for an explanation of the parameters of the model components point source (p) and Gaussian source.

	model	a			b		
	λ	8.5 μm	10.5 μm	12.5 μm	8.5 μm	10.5 μm	12.5 μm
p	f_p	0.57 ± 0.14	0.51 ± 0.12	0.53 ± 0.13	0.59 ± 0.14	0.50 ± 0.12	0.53 ± 0.13
	F_p [Jy]	0.51 ± 0.13	0.40 ± 0.09	0.97 ± 0.24	0.52 ± 0.13	0.40 ± 0.09	0.97 ± 0.24
Gauss	f_g	0.43 ± 0.10	0.49 ± 0.12	0.47 ± 0.11	0.41 ± 0.10	0.50 ± 0.12	0.47 ± 0.11
	F_g [Jy]	0.39 ± 0.09	0.39 ± 0.09	0.86 ± 0.20	0.37 ± 0.09	0.40 ± 0.09	0.86 ± 0.20
	Δx	(0)	(0)	(0)	26.0 ± 0.2	4.9 ± 1.7	5.2 ± 2.5
	Δy	(0)	(0)	(0)	23.4 ± 0.4	20.7 ± 3.3	25.5 ± 4.7
	Θ_g	($\gtrsim 40$)	($\gtrsim 50$)	($\gtrsim 60$)	9.6 ± 0.4	20.7 ± 3.3	31.4 ± 1.9
	# DOF	15			12		
	χ^2	348	84	134	169	51	104
	χ_r^2	23.2	5.6	8.9	14.0	4.2	8.7

Since the visibilities do not show a radial drop-off, we cannot constrain Θ_g with this model, but only give a *lower* limit that corresponds to a fully resolved Gaussian component at the shortest baselines. Note that this limit increases from ≈ 40 mas at 8.5 μm to 60 mas $\approx 12.5/8.5 \cdot 40$ mas at 12.5 μm, i.e. the size is less constrained at 12.5 μm because the resolution is lower. Within the errors, the point source and the over-resolved Gaussian contribute the same to the total flux at all wavelengths. With 15 degrees of freedom (N_{free}), this model yields $\chi_r^2 = \chi^2/N_{\text{free}}$ values of 23.2, 5.6 and 8.9 for the 8.5 μm, 10.5 μm, 12.5 μm fit respectively. Following tests of systematic errors and a careful treatment of the statistical errors, this model must be excluded. In other words: the model residuals are much larger than any reasonable estimate of the uncertainties of the data. A more complex model is needed.

Model b: Point source + offset circular Gaussian In model b the same components as in model a are employed, but we no longer fix the center of the Gaussian component to $(x, y) = (0, 0)$. Due to the lack of absolute astrometry and phase information, we can in fact only constrain an offset between the two sources and not their absolute positions. This implies that the source positions are interchangeable. We therefore give *offsets* $(\Delta x, \Delta y)$ *between* the components rather than their *positions*. See Figure 3.7 for a sketch of the model.

3. *Centaurus A: Dissecting the nuclear mid-infrared emission in a radio-galaxy*

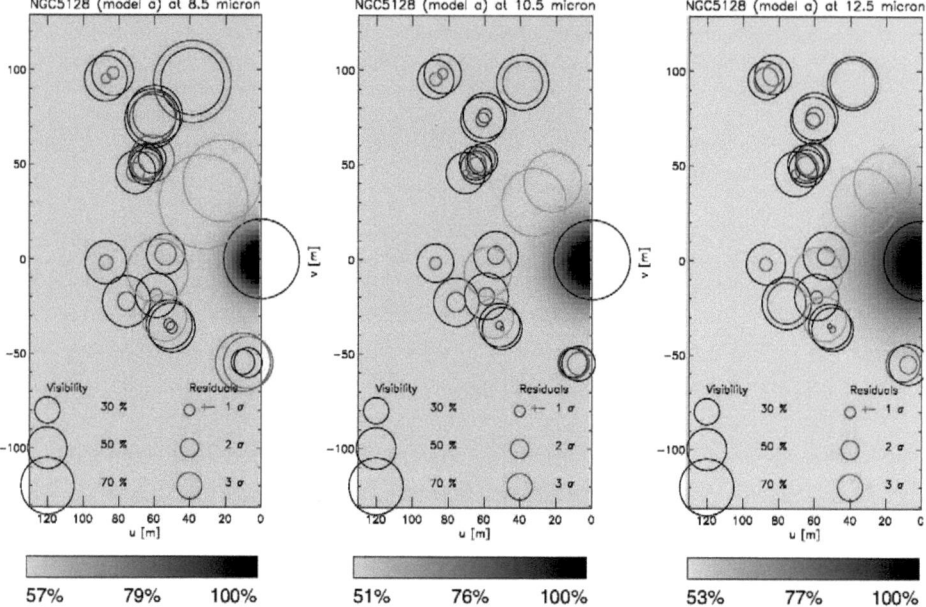

Figure 3.8.: *Model a:* Visibilities, averaged at 8.5 ± 0.2 μm, 10.5 ± 0.2 μm and 12.5 ± 0.2 μm in the (u,v) plane. Each circle (green: 2005, black: 2008) represents a visibility measurement; the radius denotes the visibility amplitude (see legend in plot). The large circle at the origin stands for the single-dish flux that corresponds to $V = 1$. The statistical errors of these visibilities are given in Table 3.2 and are not shown in the plot for clarity. Concentric to each black circle is a second circle whose radius denotes the model residual $|V_{\mathrm{obs}} - V_{\mathrm{model}}|$ (in units of σ, see legend in plot) at this point in (u,v) space. If $V_{\mathrm{obs}} - V_{\mathrm{model}} > 0$, this circle is plotted in blue, otherwise in red. The model visibilities are shown in greyscale in the backdrop image and the scaling is given below the (u,v) plane plot.
In real space, this model consists of a circularly symmetric Gaussian and a concentric, unresolved point source. The best fitting parameters of this model are given in Table 3.3.
Note that only half of the (u,v) plane is shown (the other half is point-symmetric to the one shown).

3.4. Modelling

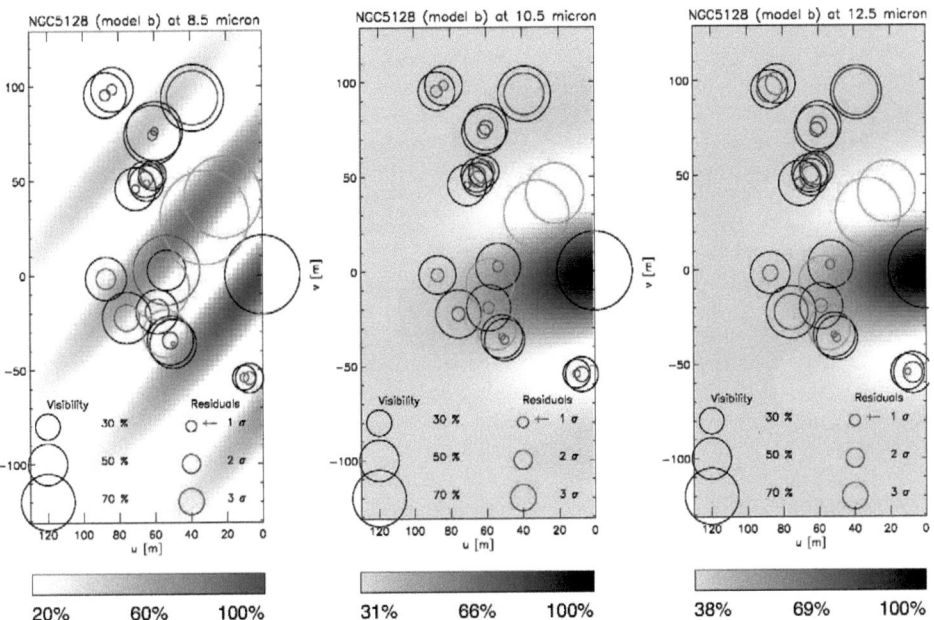

Figure 3.9.: *Model b:* Visibility model and residuals plot in the (u, v) plane (see the caption of Figure 3.8 for details).

In real space, this model consists of an unresolved point source and a circularly symmetric Gaussian that is offset from the point source by 35, 21, 26 mas at a position angle of 45°, 13°, 12° at 8.5 μm, 10.5 μm, 12.5 μm, respectively. The best fitting parameters of this model are given in Table 3.3.

3. Centaurus A: Dissecting the nuclear mid-infrared emission in a radio-galaxy

The best-fitting parameters for this model are given in Table 3.3 and the (u, v) plane fit is visualized in Figure 3.9.

For the 8.5 μm visibilities, the best-fitting offset is $(\Delta x, \Delta y) = (26.0, 23.4)$ mas or $\sqrt{\Delta x^2 + \Delta y^2} \approx 35$ mas at a position angle $\tan^{-1}(\Delta x/\Delta y) \approx 45°$ east of north. This offset matches well the co-sinusoidal pattern described before but the fringe contrast of that pattern decreases too fast because the Gaussian is constrained to be large due to the low visibilities at short baselines. At other wavelengths, this pattern is not seen so strongly and the best fit is found for smaller offsets of 21 and 26 mas and for smaller position angles of 13° and 12° for the 10.5 μm and 12.5 μm fits, respectively. The sizes of the Gaussians range from a marginally resolved one at 8.5 μm to almost over-resolved ones at 10.5 μm and 12.5 μm. Within the errors, the point source and the over-resolved Gaussian contribute equally to the total flux at all wavelengths.

This model yields a significantly reduced χ^2 value of 14.0 at 8.5 μm but is only a minor improvement at the other two wavelengths.

Apparently, allowing an offset between the point source and the circular symmetric Gaussian does not help since this model is torn between an under- and an over-resolved Gaussian: The first is needed to produce high visibilities ($\gtrsim 80\%$ at 8.5 μm) at $(u, v) \approx (40, 100)$, the latter for shallow fringe contrasts ($\lesssim 30\%$ at 8.5 μm) at $(u, v) \approx (10, -50)$ (see Figure 3.9). This cannot work and large residuals remain. The fact that the best-fit offset changes so drastically with wavelength is hard to explain physically and most probably an indication that this offset is not real.

Model c: Point source + concentric elongated Gaussian Model c explores the possibility of an elongated structure. It consists of a point source and a concentric elongated Gaussian, i.e. it is the model discussed by Meisenheimer et al. (2007). The elongated Gaussian is described by its axis ratio ρ, the FWHM of its *minor* axis Θ_e and the position angle PA (east of north) of the *major* axis (see Figure 3.7 for a sketch of the model). In fact, we cannot constrain ρ and Θ_e in any of the fits and only their product, i.e. the FWHM of the major axis is a meaningful number.

The best-fitting parameters for this model are given in Table 3.4 and the (u, v) plane fit is visualized in Figure 3.10.

In this model, the major axis is $> \lambda/2B$, i.e. over-resolved at the shortest baselines, and the minor axis is found to be point-like and responsible for the visibility variations in the north-eastern quadrant of the (u, v) plane. In effect, this fit describes a narrow "bar" in (u, v) space (see Figure 3.10) where visibilities are ≈ 1. Outside of the bar they drop quickly to f_p. The best fitting position angle of this structure is 119° (at all wavelengths) in real space which translates to 119°-90° = 29° in (u, v) space. Within the errors, the point source and the over-resolved Gaussian again contribute the same to the total flux at all wavelengths. The PA is not well constrained in this model. In fact, Burtscher et al. (2010) found a fit to the 12.5 μm visibilities with very similar χ_r^2 for a position angle of 15°.

Compared to model b, this model yields another significant reduction of χ_r^2 at 8.5 μm, but only a minor improvement at 12.5 μm and no change at 10.5 μm.

3.4. Modelling

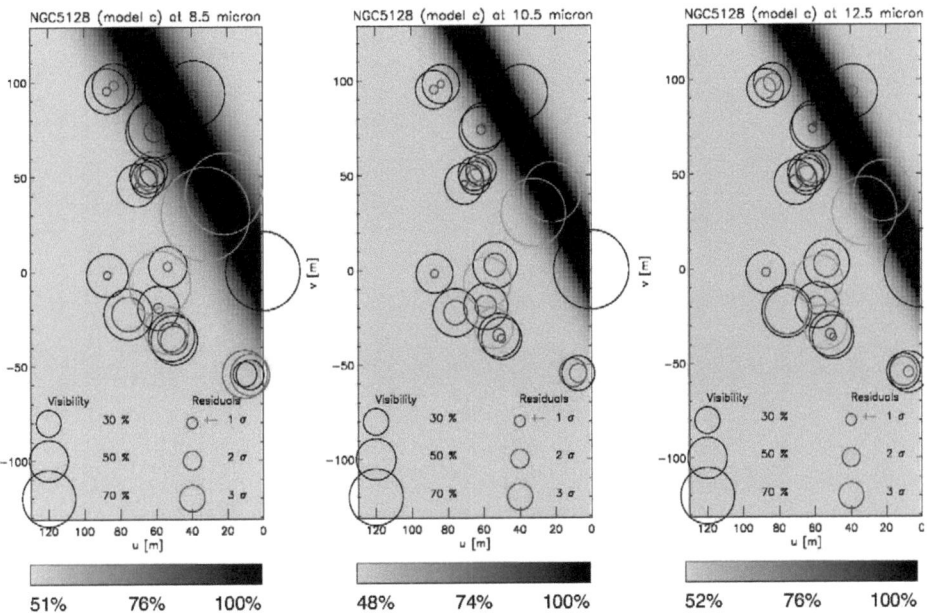

Figure 3.10.: *Model c:* Visibility model and residuals plot in the (u, v) plane (see the caption of Figure 3.8 for details).
In real space, this model consists of an unresolved point source and a concentric elongated Gaussian. The position angle of the elongated Gaussian is 119°; its minor axis is point-like and the major axis is nearly over-resolved. This results in a narrow bar in the (u, v) plane at a position angle of $\approx 29°$. The best fitting parameters of this model are given in Table 3.4.

3. Centaurus A: Dissecting the nuclear mid-infrared emission in a radio-galaxy

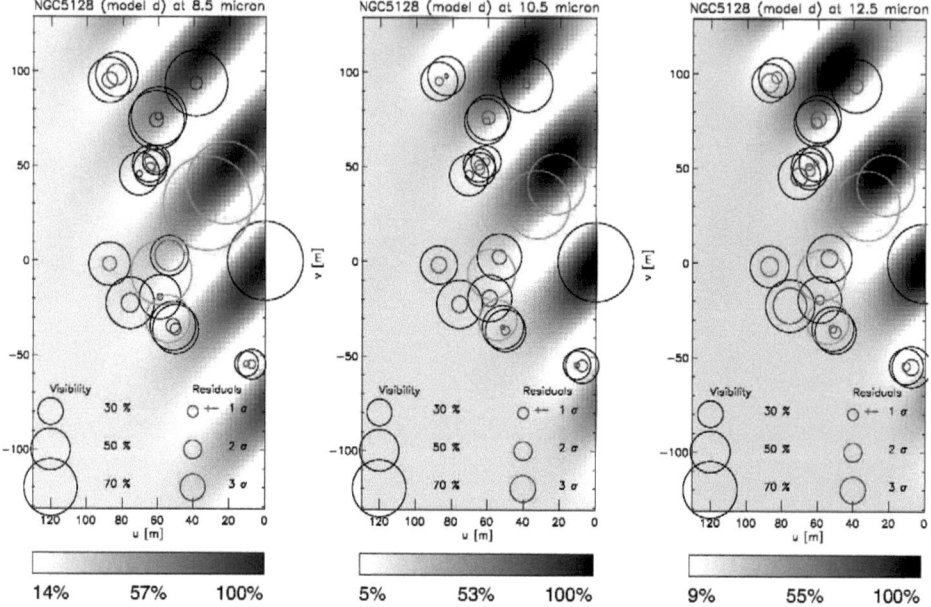

Figure 3.11.: *Model d:* Visibility model and residuals plot in the (u, v) plane (see the caption of Figure 3.8 for details).

In real space, this model consists of an unresolved point source and an elongated Gaussian that is offset from the point source by 34, 40 and 47 mas at 47°, 44° and 43° for the 8.5 μm, 10.5 μm and 12.5 μm fits respectively. The position angle of the elongated Gaussian is $\approx$ 110°; its minor axis is point-like and the major axis is nearly over-resolved. Together with the offset point source, this results in a co-sinusoidal pattern in the (u, v) plane at a position angle of $\approx$ 20°. The fringe contrast in that direction does not decrease because of the point-like minor axis; in the orthogonal direction it decreases rapidly due to the relatively large major axis of the elongated Gaussian. Note that the visibility decreases to very low values where the offset point source interferes destructively with the elongated Gaussian component. The best fitting parameters of this model are given in Table 3.4.

3.4. Modelling

Table 3.4.: Same as Table 3.3, but for the models c and d. See the text for an explanation of the parameters of the model components point source (p) and elongated Gaussian source.

	model	c			d		
	λ	8.5 μm	10.5 μm	12.5 μm	8.5 μm	10.5 μm	12.5 μm
p	f_p	0.51 ± 0.12	0.48 ± 0.12	0.52 ± 0.13	0.57 ± 0.14	0.52 ± 0.13	0.55 ± 0.13
	F_p [Jy]	0.46 ± 0.11	0.38 ± 0.09	0.95 ± 0.24	0.51 ± 0.13	0.41 ± 0.10	1.00 ± 0.24
Elong. Gauss	f_e	0.49 ± 0.12	0.52 ± 0.13	0.48 ± 0.12	0.43 ± 0.11	0.48 ± 0.12	0.45 ± 0.11
	F_e [Jy]	0.44 ± 0.11	0.41 ± 0.10	0.88 ± 0.22	0.39 ± 0.10	0.38 ± 0.09	0.82 ± 0.20
	Δx	(0)	(0)	(0)	25.0 ± 0.7	27.7 ± 1.3	32.2 ± 1.6
	Δy	(0)	(0)	(0)	23.4 ± 0.5	28.6 ± 1.2	34.3 ± 0.7
	PA_e	119.2 ± 0.4	118.8 ± 1.3	119.2 ± 1.7	105.7 ± 3.6	111.7 ± 4.3	115.5 ± 1.5
	$\Theta_e \cdot \rho_e$	≈60	≈100	≈130	≈22	≈35	≈48
	# DOF	12			10		
	χ^2	141	51	89	46	16	41
	χ_r^2	11.7	4.2	7.4	4.6	1.6	4.1

Model d: Point source + offset elongated Gaussian Finally, we allow an offset $(\Delta x, \Delta y)$ between the point source and the elongated Gaussian of model c (see Figure 3.7 for a sketch of the model). For the offset, the same restrictions apply as mentioned above for model b.

This model has two almost equally good (in terms of χ_r^2) solutions for very different values of $(\Delta x, \Delta y)$. The formally best model (with $\chi_r^2 = 3.3, 1.4, 4.0$) is found for offsets of $(\Delta x, \Delta y) \approx (20, -12), (30, -16), (41, -24)$ at 8.5 μm, 10.5 μm, 12.5 μm, respectively. The second best model (with $\chi_r^2 = 4.6, 1.6, 4.1$) is found for offsets of $(\Delta x, \Delta y) \approx (25, 23), (28, 29), (32, 34)$. The other fitted parameters $(f_g, f_e, PA, \Theta_e, \rho_e)$ are similar for the two models. The ambiguity in $(\Delta x, \Delta y)$ can be most clearly seen in the respective χ^2 planes, shown in Figure 3.12: At 8.5 μm, actually three minima can be seen (the ones given above and an additional one at $\approx (20, -45)$).[8] The latter minimum is not seen at other wavelengths and therefore rejected.

With regard to the width of the χ_r^2 distribution for this model, $\sigma_{\chi_r^2} = \sqrt{2/N_{\text{free}}} = 0.45$,

---

[8]In Figure 3.12, the $(\Delta x, \Delta y)$ landscape is plotted only for $\Delta x = [0,50]$ mas and $\Delta y = $ [-50, 50] mas, since it repeats for larger offsets, see Figure 3.13 for a demonstration at 10.5 μm. For this model, the distance between the maxima of the fringe pattern in the (u, v) plane (see Figure 3.11) is set by the offset of the two components. Multiples of this offset lead to accordingly decreased distances in the (u, v) plane (cf. "harmonics of a wave"). Offsets $\gtrsim 50$ mas lead to fringe patterns that show variations on scales $\lesssim$ the telescope diameter in the (u, v) plane and are therefore likely to fit noise rather than real structure. We therefore choose the smallest offsets compatible with the data.

only at 8.5 μm a marginally significant difference can be claimed for the goodness of fit between the two offsets.

However, in the model with the formally best fit, the offset position changes drastically between the wavelengths. This is not expected for a real source. In the second best fit, the offset position also changes with wavelength, but roughly according to the wavelength ratio which can be understood as an effect of the sparsely sampled (u,v) plane.[9]

We therefore report the best fitting parameters of the fit with the more consistent $(\Delta x, \Delta y)$ values over wavelength. These parameters are given in Table 3.4 and the (u,v) plane fit is visualized in Figure 3.11. Again, the elongated Gaussian's minor axis is point-like and the major axis is close to being over-resolved at all wavelengths. This leads again to a "bar-like" structure as in model c – but now co-sinusoidally modulated due to the offset of the two source components. The offsets found in this fit correspond to $\sqrt{(\Delta x)^2 + (\Delta y)^2} \approx 34$, 40 and 47 mas at 47°, 44° and 43° for the 8.5 μm, 10.5 μm and 12.5 μm fits respectively. The position angle of the elongated Gaussian is smaller than in model c. Within the errors, the point source and the over-resolved Gaussian contribute equally to the total flux at all wavelengths.

This model, that has lost two degrees of freedom in comparison with model c, is a significant improvement in terms of χ_r^2 at all wavelengths, leading to acceptable values of $\chi_r^2 = 4.6$, 1.6, 4.1 at 8.5 μm, 10.5 μm and 12.5 μm respectively.

Cuts through the χ^2 space (Figure 3.12) demonstrate the parameter constraints: The position angle and the relative flux levels of the point source and the elongated Gaussian are well defined. Regarding ρ and Θ_e, only their product is constrained and the $(\Delta x, \Delta y)$ χ^2 plane is complex.

---

[9]If the offsets were identical at all wavelengths, then the true maximum of the fringe pattern produced by the offset source (see Figure 3.11) would be at the same spatial frequencies at all wavelengths. In a (u,v) plot in units of meters (such as Figure 3.11), the maximum would then move to longer baselines for longer wavelengths. Due to the sparsely sampled (u,v) plane, this cannot be excluded by the data: It is quite possible that the true maximum is only seen at 8.5 μm (at $(u,v) \approx (50,80)$), where the visibilities are highest, and that at other wavelengths only the wings of this maximum are seen. The errors of $(\Delta x, \Delta y)$ only describe the statistical uncertainties of the offset position. The systematic uncertainties due to the asymmetric and sparse (u,v) coverage are larger but hard to quantify. For the same reason, the derived position angle must be taken with care.

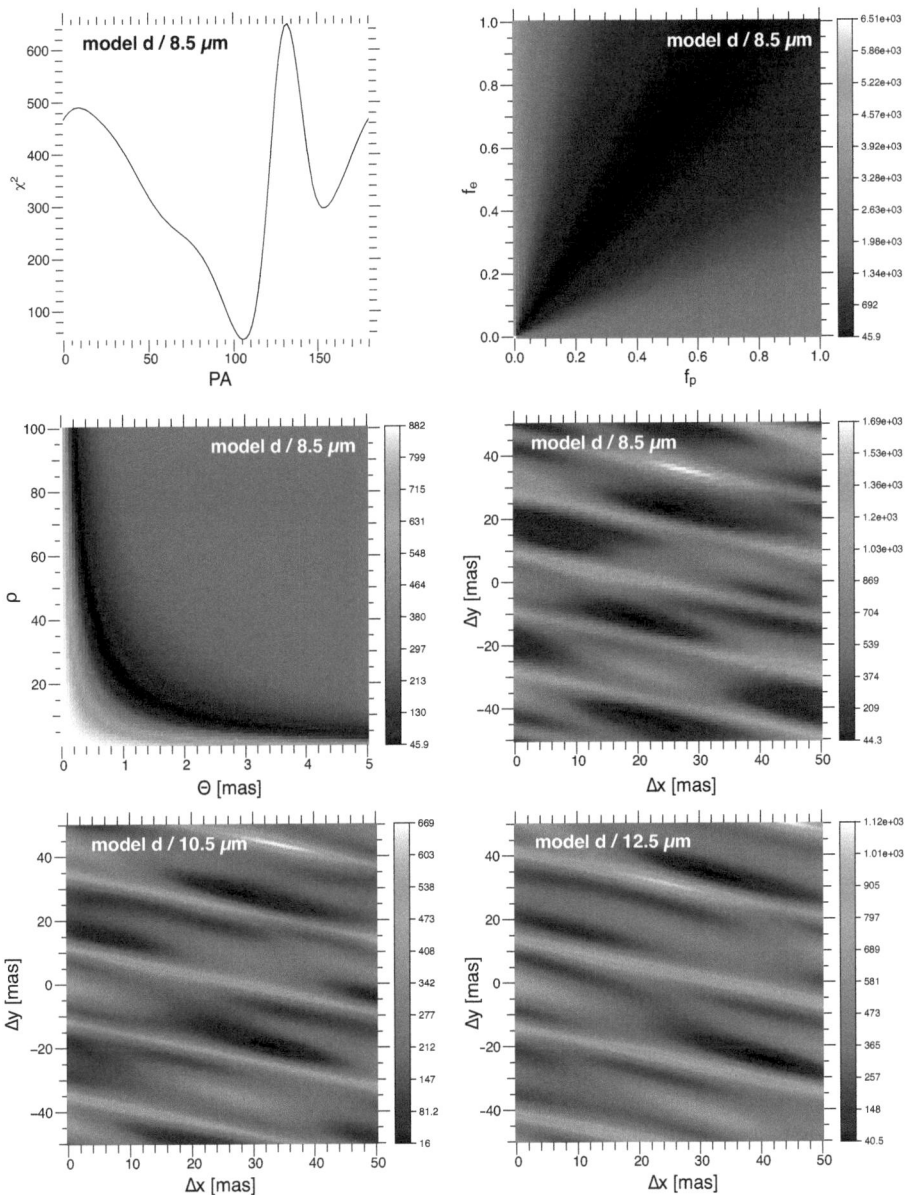

Figure 3.12.: Cuts through the χ^2 space for model d.

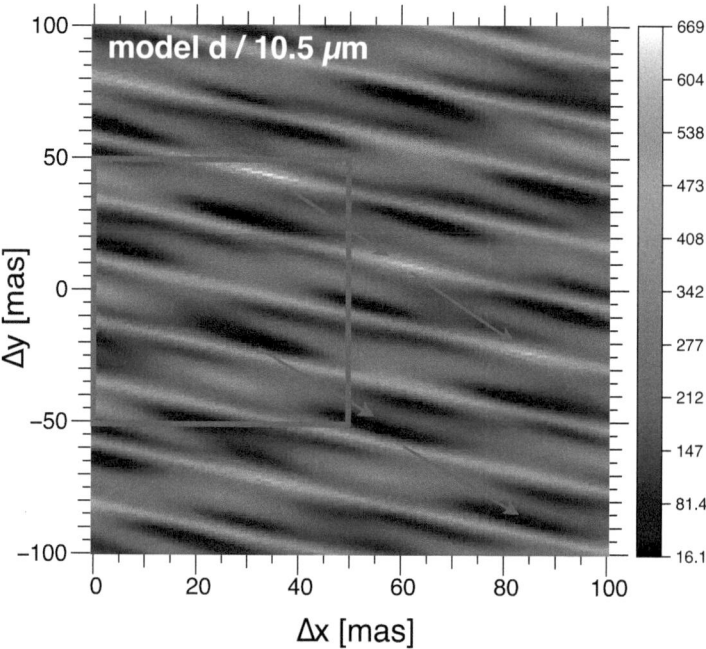

Figure 3.13.: Cut through the χ^2 space for model d at 10.5 μm with a larger plot range compared to Figure 3.12, showing that the χ^2 pattern is repeated beyond $(\Delta x, \Delta y) \approx (50, 50)$ mas (red box). Examples for a repeated maximum and for a repeated minimum in the χ^2 plane are illustrated with arrows.

3.5. Discussion

The discussion is structured as follows: First, a critical discussion of the model fits is given and the spectra of the source components are briefly discussed. Next, variability is studied. Taken together, astrophysical scenarios are sketched. At the end, we compare these observations to other MIDI observations of AGNs.

3.5.1. Geometrical model fits

Summary One of the most crucial aspects of χ^2 fitting is the correct determination of the statistical errors. Only with realistically estimated uncertainties can the value of χ_r^2 be used to estimate the quality of the fit. For the data modeled here, the errors have been computed taking into account the errors of the individual observations and the uncertainties of the calibration templates. By repeating an observation we made sure that systematic errors do not dominate over the statistical ones.

In conclusion, models a (concentric point source and circularly symmetric Gaussian), b (offset point source and circularly symmetric Gaussian) and c (concentric point source and elongated Gaussian) must be rejected because no fit could be found that results in a χ_r^2 near the desired value of 1.

This leaves us with model d, the point source and offset elongated Gaussian. It has two equally good solutions for $(\Delta x, \Delta y)$. In both, the offset between the two sources differs significantly from wavelength to wavelength. While the marginally better fit has offset differences that cannot be explained, the second solution's offset differences can be explained as effects arising from the sparse (u, v) sampling. Model d, with the parameter values reported in Table 3.4, is therefore our best geometrical explanation for the visibility pattern.

It is worth noting that in all models the best fit is achieved at 10.5±0.2 μm. The reason for this is simple: At this wavelength regime, near the bottom of the pronounced silicate absorption trough, the SNR is lowest (the errors are largest) and therefore the residuals in units of σ are smallest. However, since there is no reason why the structure should be better described by our model geometry at 10.5 μm than at 8.5 μm or 12.5 μm, we conclude that either the errors at 10.5 μm are *over*-estimated by a factor of $\sqrt{\chi_r^2(8.5\mu m)/\chi_r^2(10.5\mu m)} \approx 1.7$ or the errors at 8.5 μm and 12.5 μm are *under*-estimated by the same factor.

Spectral Index of the model components So far we have concentrated on fits at three distinct wavelengths. Combining the results at these wavelengths gives an estimate of the spectral index[10] of the fitted components. The contributions of the two components of

---

[10] Note that calculating a spectral index out of a fit of a geometrical model to interferometric data can be dangerous since the resolution is higher at 8.5 μm than at 12.5 μm. Effectively we are observing with a larger aperture at 12.5 μm compared to 8.5 μm. The fitted 12.5 μm flux is thus an upper limit to the "true" 12.5 μm flux (that would be observed with a hypothetical, smaller, 8.5 μm effective aperture).

model d are consistent with both the point source and the extended source having similar spectra, e.g. a flat, $F_\nu \propto \nu^{-0.36}$, power-law spectrum (as inferred from the synchrotron model of Meisenheimer et al. (2007) for the point source) or a warm (≈ 500K) blackbody spectrum.

The visibilities show no silicate feature, because the depth of this feature is identical (within the errors) in the single-dish spectra and the correlated spectra. In the context of geometrical model d, this implies that the optical depth to dust obscuration is constant within the central ca. 50 mas (see Figure 3.16).

The 2005 observations Although the 2005 observations were not included in the fit, they are also plotted in the model images, Figure 3.8 - 3.11, for comparison. Models a and b are even stronger rejected when adding the 2005 observations. Model c was discussed in Meisenheimer et al. (2007) as a possible interpretation for the 2005 visibilities. For the following discussion, it is interesting to note that model d, optimized just for the 2008 visibilities, seems to be a relatively good model for the 2005 visibilities, too (see Figure 3.11).

3.5.2. Variability

The correlated and total flux of Cen A changed between our 2005 and 2008 observations in such a way as to leave the visibilities on the U3U4 baseline unchanged. What does this tell us about the source morphology, if we assume that model d is a valid description of the 2008 morphology?

Basically, two scenarios are possible

1. Either the *geometry did not change*, implying that both components brightened. In the context of model d, the minimum separation between the two source components is 34 mas, corresponding to about 2 light years in Cen A. Since the observations are about 3 years apart, the brightness increase in the two components can be causally connected. Also, the 2005 observations are compatible with the existence of a point source and an elongated Gaussian.

2. The other possible scenario is that only one component increased its flux, implying a *change in geometry* to keep the visibilities constant. Would that mean that one of the components moved? In this case the three years of time difference between the two epochs would require velocities near c, too fast for a moving dust cloud (the dust would be rapidly destroyed by shocks) and untypically fast for the moving knots in the jet.

In order to understand the origin of the variability, it would therefore be helpful to get an estimate of *when* the increase in flux occurred.

Variability time constraints from an IR-X-Ray correlation Since Cen A is not continually monitored in the mid-infrared, the precise date of flux increase in this waveband

3.5. Discussion

Table 3.5.: Correlation between X-Ray counts and mid IR flux for Cen A

Date	MJD	MIR flux [Jy] at 11.7 μm	XTE counts at MJD-50	IR flux / X-Ray counts Jy / (counts/s)	Reference & Notes
2002-01-30	52304	≈ 1.3	0.89	1.5	(1)
2002-06-28	52453	1.6 ± 0.2	0.78	2.1	(2)
2005-02-28	53429	1.10 ± 0.08	0.50	2.2	(3)
2006-03-15	53809	1.150 ± 0.005	0.57	2.0	(4)
2006-12-27	54096	≈ 1.1	0.58	1.9	(5)
2008-04-20	54576	1.50 ± 0.05	0.95	1.6	(3)

References: (1): Siebenmorgen et al. (2004) measured 650 mJy at 10.4 μm using the TIMMI2 mid-infrared camera at the ESO 3.6 m telescope. Using a factor of 2 (derived from our 2005 and 2008 spectra) to convert this 10.4 μm flux to an 11.7 μm flux, we arrive at the quoted 1.3 Jy.
(2): Whysong & Antonucci (2004); Keck I telescope
(3): our observations
(4): Reunanen et al. (2010); VISIR/VLT
(5): van der Wolk et al. (2010); VISIR / VLT. Their flux was given at 11.85 μm and a spectral correction factor of ≈ 0.9 was applied to estimate an 11.7 μm flux.

cannot easily be reconstructed. However, there is one observation that directly helps to constrain the date. On 2006-12-27, van der Wolk et al. (2010) used the VISIR instrument on the VLT UT3 and observed a core flux at 11.85 μm of (1200 ± 47) mJy, compatible with our 2005 photometry and significantly lower than our 2008 observation. Since the VISIR aperture is very similar to the one used for the MIDI total flux observations and since van der Wolk et al. (2010) saw Cen A essentially as a point source, we can directly compare their flux to our total flux measurement and find that the bulk of the flux increase must have occurred after van der Wolk et al. (2010)'s observation.

In 2007 there was unfortunately no observation of Cen A in the mid-infrared with a high-resolution camera (e.g. VLT / VISIR or Gemini-South / T-ReCS). In order to further constrain the date of flux increase, we therefore looked at monitoring data of the All Sky Monitor (ASM) onboard the Rossi X-Ray Timing Explorer (RXTE), see Figure 3.14. With this satellite, Cen A is monitored every day at 2-10 keV. When applying a time lag between the X-Ray counts and the IR flux of about 50 days (in the sense that the IR radiation follows the X-Ray radiation), then the IR flux appears to be correlated with the X-Ray counts since 2002 (see Table 3.5). The time lag of ≈ 50 days could be understood as the light travel time from the inner hot accretion disk (Evans et al. 2004) to the innermost radius of dust. The sublimation radius in Cen A was estimated by Meisenheimer et al. (2007) to be $\gtrsim$ 0.013 pc (≈ 13 light days).

Turning again to Figure 3.14, the most prominent feature of this lightcurve actually is the strong outburst in mid-November 2007. Taken together with the apparent correlation between X-Ray counts and IR flux, we speculate that this outburst was responsible for the increase in IR flux between the two MIDI observations. The time difference between

3. Centaurus A: Dissecting the nuclear mid-infrared emission in a radio-galaxy

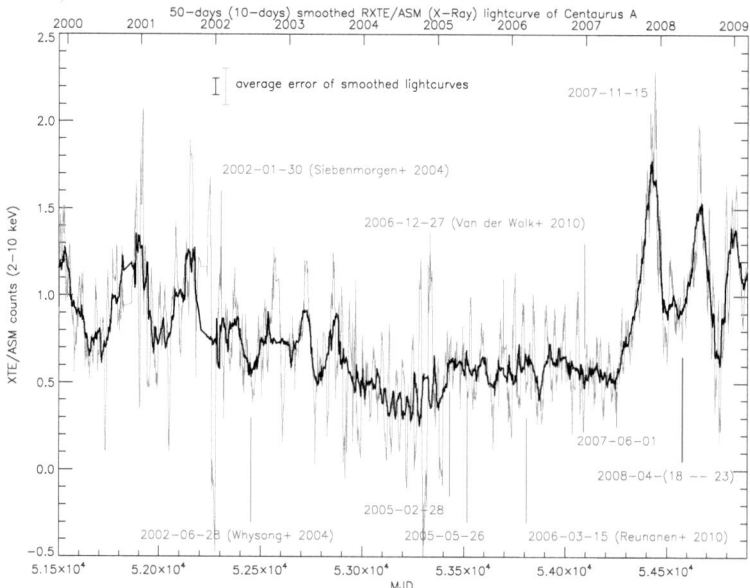

Figure 3.14.: X-Ray lightcurve from the All-Sky-Monitor (ASM) onboard the Rossi X-Ray Timing Explorer (RXTE), summed over all bands of the instrument and smoothed by 50 days (black curve) and 10 days (grey curve), respectively. The lower x axis displays the Modified Julian Date (MJD). The nights of the MIDI observations are marked. A rather quiescent X-Ray phase in 2005 is followed by a strong outburst in mid November 2007, five months before the second MIDI observations. The averaged X-Ray count rate is taken to be 0.6 (per second) at the 2005 observations and 0.9 in April 2008. For the other marked dates, see Table 3.5 and text.

the X-Ray flare[11] and our 2008 observations is then $\lesssim 8$ months.

VLBI monitoring For Cen A, correlations of flares in hard X-Rays with the appearance of new knots in the radio jet have been reported by Tingay et al. (1998). It is therefore tempting to look for new components that might have followed the X-Ray outburst in November 2007 and that could explain the increase in mid-IR flux. Tingay et al. (1998) also observed short-term ($\approx$ months-scale) variability of the jet components in an 8.4 GHz

---
[11]Taking the half-rise date, i.e. when the X-Ray counts first reached the high level.

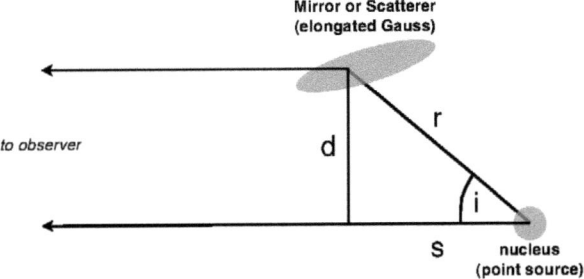

Figure 3.15.: Derivation of the position of the elongated Gaussian, see text for explanation

light curve. Can we find a radio counterpart to the components of model d and did the radio morphology change between our two observations?

Although Cen A is very frequently observed using the highest resolution radio technique, Very Long Baseline Interferometry (VLBI), there have been no observations between the extensive monitoring campaign of Tingay et al. (2001) that ended in January 2000 and Tingay & Lenc (2009) who published a 2.3 GHz map of June 2007, i.e. five months before the X-Ray flare (Steven Tingay, priv. comm.). Müller et al. (2010b) and Müller (2010) report a series of VLBI observations at 8.4 GHz from the time of the X-Ray flare in late 2007 until about a year after. Comparing VLBI maps between 1997 and 2008, they find no new (bright) components (see Figure 5.24 of Müller (2010)), but they caution that due to different beams and sensitivities it is difficult to compare the observations. Still, this leaves little room for the appearance of a new component between 2005 and 2008. Furthermore, if a new component had been ejected at the X-Ray outburst in Nov 2007, it would be unlikely that it had already reached a separation of $\approx$ 30 mas in April 2008, since the bulk motion of the jet is only 0.5 c (8 mas/yr) and proper motion of the components is even slower at $\approx$ 2 mas/yr, (Tingay et al. 2001).

A projection effect? *Movement of material* from the nucleus to the offset position found in model d cannot have occurred in less than 8 months time without employing extreme apparent superluminal motions that have not been observed for Cen A so far.

Therefore, it seems reasonable to assume that the mid-IR emission geometry did not change between 2005 and 2008 (and that model d is a valid description for both epochs). We further assumed (see above) that the IR outburst was *triggered* by the X-Ray flare in November 2007. The flux increase then occurred first in what we model as point source and then gets reflected (scattered or re-emitted) by what we model as elongated Gaussian. In this case, we can constrain the angle i between our line of sight and the elongated Gaussian's line of sight to the nucleus by using the geometrical relations in a triangle with known sides, see Figure 3.15.

We have measured d to be $\gtrsim 34$ mas (from model d at 8.5 μm, i.e. at highest resolution) and we have a constraint from the X-Ray monitoring for $r - s \lesssim c \cdot 8$ months (taking the half-rise time of the X-ray flare) ≈ 10 mas. This allows us to estimate i since $(r - s)/d = (1 - \cos i)/\sin i \lesssim 10/34$ or $i \lesssim 40°$.

3.5.3. The elongated source and overall geometry

From our geometrical model fits, only the offset between the elongated source and the unresolved point source is determined. The offsets found were compatible with the source lying either in the north-east or in the south-west.

However, taking the variability considerations into account, the elongated source must be located "in front of" the point source for the projection effect to work. Due to the history of non-thermal flux variations in Cen A and the inferred (projected) proximity of the elongated Gaussian structure to the jet, it is likely that the extended component is actually connected with the jet region. In this case, the structure must be to the north-east of the nucleus since that part of the jet points in our direction. In fact the inclination implied from the variability consideration ($i \lesssim 40°$) is in between the different inclinations inferred from radio and X-ray studies: In VLBI observations, Tingay et al. (2001) found an inclination (angle to the line of sight) of $50° - 80°$ (Tingay et al. 1998) but in combined radio and X-ray observations, Hardcastle et al. (2003b) find that only small inclinations of $\approx 20°$ are consistent with both X-rays and radio observations from combined constraints on apparent motions and sidedness of the jet (components).

We note that it is unlikely that the elongated structure is connected to the nuclear disks found by Neumayer et al. (2007) and Espada et al. (2009) in molecular hydrogen and CO emission, respectively.

Our best model scenario, together with the VLBI radio data is sketched in Figure 3.16.

3.5.4. Comparison with other MIDI AGN observations

In the two other AGNs that have been studied extensively with MIDI, the Circinus galaxy and NGC 1068, "scatter" of the visibilities in the (u, v) plane has only been seen at very low visibility levels ($\lesssim 20$ %) where the source is almost completely resolved (cf. Figure 4 of Tristram et al. (2007)). In the Circinus galaxy, a fit with two Gaussian components leads to a $\chi^2 \sim 16600$ with 451 degrees of freedom (Tristram et al. 2007); in NGC 1068 especially the data at longer baselines are also not well described by a model of two Gaussian components (Raban et al. 2009). These small scale structures can, for example, arise from clumpiness in the dusty disk as has been demonstrated by Tristram et al. (2007).

It is easily understandable that deviations from the usual Gaussian fits only arise at low visibilities because at large visibilities most simple geometries (Gaussians, disks, rings, ...) lead to $V \propto (BL/\lambda)^{-2}$, see Section 2.2.6. Subtracting the point source flux fraction in Cen A ($\approx 50\%$) from the visibilities, leaves also only very low visibility values. Therefore, we are effectively observing the extended component in Cen A at visibilities $\lesssim 20$ % or, equivalently, probing the extended component at relatively high spatial resolutions which are most sensitive to small-scale structures and not well described by smooth structures. This is the explanation why the smooth fits (Gaussian, elongated Gaussian) cannot explain the visibility pattern: Large, smooth structures are probably only good estimates for the real surface brightness distribution at large visibilities (there, they describe the envelope of the true, probably more complicated, visibility function). Taking into account that the nuclear region of Cen A, especially around the jet, is very complex (Neumayer et al. 2007), it is actually not too surprising that no perfect fit can be found with simple model components.

3.6. Conclusions

- The simplest geometrical models (a,b,c) cannot explain the complex visibility pattern of Cen A. After having studied systematic errors, including calibration errors and propagating them to the final visibilities, we are convinced that the residuals from these models are not caused by underestimated errors.

- Acceptable fits are achieved at 8.5 μm, 10.5 μm and 12.5 μm for the elongated Gaussian + offset point source model. There are two solutions for the offset of which one was discarded because the offset changed excessively with wavelength. In the chosen solution the offset between the two components is 34, 40 and 47 mas for the three wavelengths respectively. In this fit, the major axis of the elongated Gaussian is at a position angle of $\approx 110°$ and is in fact almost over-resolved with a FWHM of 22 - 48 mas; the minor axis is unresolved. Both components contribute roughly the same amount of flux. In this model, the remaining scatter can be explained because such a simple structure (especially the extended component with unresolved and almost over-resolved axes) is obviously not a very good fit for the complex nuclear region of Cen A, especially close to the jet. However, more complex geometrical models lead to a very low number of degrees of freedom.

- The nucleus of Cen A brightened considerably between the 2005 and 2008 MIDI observations. Repeated observations on one baseline showed that the visibilities did not change between these epochs suggesting that the mid-IR emitting geometry stayed the same. An alternative explanation that is compatible with the datasets from both epochs would be that a component newly appeared in 2008 (where the model with the offset component results in a good fit). This scenario seems to be excluded by X-Ray and radio observations.

- Assuming that the mid-IR geometry is well explained by the offset two-component model and indeed did not change between 2005 and 2008, leads to the question *when* the two components brightened. A mid-IR observation at the end of 2006 showed no increase in flux relative to our 2005 observation. In November 2007, a bright X-Ray flare was detected in Cen A suggesting that this could be responsible

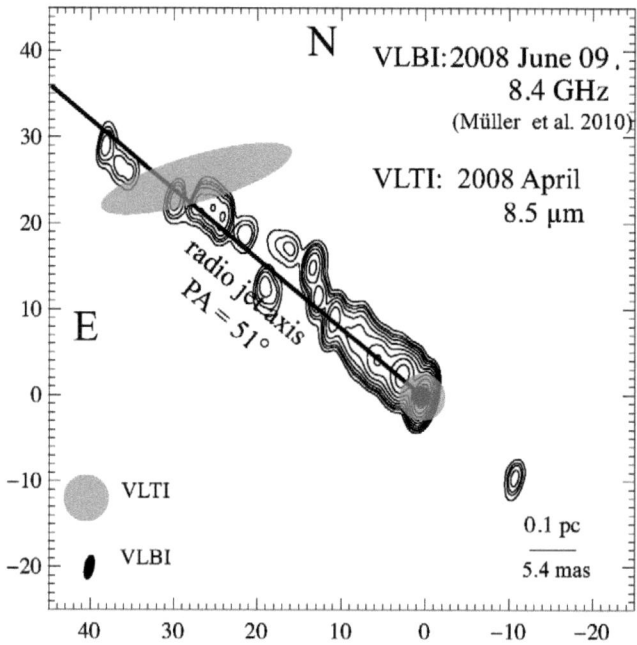

Figure 3.16.: Best-fitting two-component geometrical model d (at the highest resolution, i.e. 8.5 μm) for the nuclear mid-infrared emission of Centaurus A consisting of an elongated Gaussian disk and an unresolved point source. The major axis of the disk is ≈ 22 mas or ≈ 0.4 pc, the minor axis of the disk is smaller than the VLTI resolution of about 5 mas at 8.5 μm. The position angle of the disk is found to be $105.7 \pm 3.6°$ in this model. The yellow circle is the point source. It has previously been identified with the synchrotron core (grey point inside the yellow circle) by Meisenheimer et al. (2007). The radio radio jet axis is at a PA $\sim 51°$ (e.g. Tingay et al. 2001) and the radio contours are from Müller et al. (2010b). The PA of the X-ray jet is $\sim 55°$ (Kraft et al. 2000). The relative positioning of the VLTI and VLBI structures follows from the variability scenario (see text).

for the outburst. In a possible scenario, the X-Ray outburst is caused by an event in the very center of Cen A which we identify with our point source. Using a light travel time argument, the extended component must then be positioned *in front of* the point source which puts it near the jet of Cen A. The nature of the extended component is not constrained by our models. However, taking its proximity to the jet into account, it is likely that it is connected to the jet flow.

3.7. Outlook

Fitting procedure The employed fit (using LITpro) was not a global grid-search (but we tried to scan the parameter space manually as good as possible). As the model χ^2 planes turned out to be highly complex (see Figure 3.12), a more robust fitting algorithm is needed.

A Monte Carlo fitting algorithm would allow us to select models with greater confidence. It would help to separate more clearly the errors arising from single-dish and from interferometric measurements, such as demonstrated in Chapter 4 for the Seyfert 1 galaxy NGC 4151 where the single-dish error was several times larger than the correlated flux errors.

A spectro-interferometric model might be able to disentangle the model components more clearly and determine their spectra. However, with the current uncertainties in the geometry of the structure, it is hard to imagine a robust solution to a spectro-interferometric model.

(u, v) **coverage** The most promising way to better constrain the source brightness distribution is to better sample the (u, v) plane. Especially short baselines are missing at the moment to discriminate models with over-resolved sources from models with offsets.

Observations with very short baselines ($\lesssim 40$ m) require the use of the flexibly positionable ATs of the VLTI. Such a study is currently under consideration. Due to their smaller mirror diameter in comparison with the larger UTs ($D_{AT}^2/D_{UT}^2 \approx 1/16$) and their poorer optical properties, these observations will be very challenging. The correlated fluxes of Cen A at short baselines can be expected to be $\lesssim 1.5$ Jy (at 12.5 μm; the level of the total flux in bright state). This "AT flux" corresponds to a "UT flux" of $\lesssim 100$ mJy. With an improved online data system at the VLTI, such observations might be just so possible, but it probably needs to be paired with new UT observations to control the longer baselines.

Such observations could provide important clues on the nature of the emission. For example, the geometrical models *c* and *d* should be easily discernible with the new observations: If there really is an offset point source (model *d*) then the visibility should decrease considerably below 50% at baselines shorter than 20m at position angles near 40°, but stay at high values near 130° out to 30m (compare Figures 3.10 and 3.11).

3. Centaurus A: Dissecting the nuclear mid-infrared emission in a radio-galaxy

Variability In the context of multi-wavelength monitoring of Cen A it would be important to determine the nuclear flux from Cen A by mid-IR interferometry with as few (u,v) points as possible. To make this possible, it is first necessary to have a reliable model of the mid-IR visibilities.

4. NGC 4151: The first resolved nuclear dust in a type 1 AGN[1]

4.1. Introduction

MIDI observations have been successful in testing the unified model of active galaxies by resolving warm dust in their nuclei (see Section 1.4). So far, however, no Sy 1s have been observed in sufficient detail[2] to test the central premise of the unified models: type 1 and type 2 dust distributions are identical and the observed differences are only due to differences in orientation with respect to the line of sight.

At a distance $D = (14 \pm 1)$ Mpc (i.e. 1 mas ≈ 0.068 pc)[3], NGC 4151 (Figure 4.1) is the closest and brightest type 1 galaxy (classification: Seyfert 1.5) – at a declination of $\approx +40°$, which puts constraints on its observability from Paranal. It is also one of the most variable Seyfert galaxies: The UV continuum flux varies on scales of days and weeks (Ulrich 2000) and the reverberation time to the hot dust on the sub-parsec scale varies on yearly timescales (Koshida et al. 2009).

Sy 1 galaxies have been observed previously with MIDI (NGC 3783, Beckert et al. 2008), (NGC 7469, Tristram et al. 2009), see also Kishimoto et al. (2009), but NGC 4151 is the first multi-baseline case where the size of the nuclear dust distribution is clearly indicated.

NGC 4151 was also the first extragalactic target to be observed successfully with optical interferometry: Swain et al. (2003) reported near-IR, 2.2μm, observations with the Keck interferometer. They find that the majority of the K band emission comes from a largely unresolved source of ≤ 0.1 pc in diameter. Based on this small size, they argued that the K band emission arises in the central hot accretion disk. They note, however, that their result is also consistent with very hot dust at the sublimation radius. This view is supported by the K band reverberation measurements of Minezaki et al. (2004) who find a lag time corresponding to ≈ 0.04 pc.

In general, reverberation-based radii r_rev were found to be *systematically* smaller by a factor of ≈ 3 than the predicted dust sublimation radii r_sub (Kishimoto et al. 2007). On the other hand, interferometrically determined radii seem to be roughly equal or only slightly larger than r_rev (Kishimoto et al. 2009).

---

[1]adapted from Burtscher et al. (2009)

[2]Since the space density of Seyfert 2 galaxies is larger than that of Seyfert 1s (Maiolino & Rieke (1995) estimate a factor of 4), there are more nearby and bright type 2 galaxies.

[3]from the NASA Extragalactic Database: http://nedwww.ipac.caltech.edu; distance from redshift with H_0=73 km/s/Mpc; other estimates range to 20 Mpc

97

4. NGC 4151: The first resolved nuclear dust in a type 1 AGN

Figure 4.1.: SDSS *gri* image of NGC 4151. The active nucleus is easily spotted. The central part of the galaxy is dominated by a "fat bar" (Ulrich 2000). The weak spiral arms can be seen in blue extending almost to the top and bottom of the image. The image is 13×9.5 arc minutes (53×39 kpc) wide. Five arc minutes to the NE, group member NGC 4156 can be seen. Credit: David W. Hogg, Michael R. Blanton, and the Sloan Digital Sky Survey Collaboration

Reverberation techniques have also been used by Bentz et al. (2006) to derive an estimate for the mass of the supermassive black hole: $M_{BH} = \left(4.57^{+0.57}_{-0.47}\right) \times 10^7 M_\odot$.

Riffel et al. (2009) modeled the near-infrared spectrum of this source and found it to be composed of a powerlaw accretion-disk spectrum and a component likely arising from hot dust (T = 1285 K). The hot dust component dominates at $\lambda \gtrsim 1.3\mu m$, consistent with the interpretation of hot dust emitting at $2\mu m$. They measured a K band flux of ≈ 65 mJy.

An early attempt to resolve the nucleus in the mid-IR is reported by Neugebauer et al. (1990). They observed NGC 4151 at 11.2 μm using the f/70 Cassegrain focus of the 200 inch (5 m) Hale telescope. They claim to have determined the size of the resolved emitter at $11.2\mu m$ to $(0.16 \pm 0.04)''$ – less than a third of the diffraction limit of the telescope. After careful study of their paper and references therein and after discussions with researchers familiar with similar high-resolution techniques (Ch. Leinert, priv. comm.), we decided not to take the resolved scale reported by Neugebauer et al. (1990) into account for our further discussion.

This is corroborated by high-resolution observations by Soifer et al. (2003) who studied the nucleus with the Keck I telescope with a measured PSF FWHM of $0.36''$ at 12.5 μm and found it unresolved. Radomski et al. (2003) presented images of NGC 4151 at $10.8\mu m$ with the Gemini North telescope (aperture: $4.5''$, measured PSF FWHM: $\approx 0.55''$). They find that the majority (73%) of the N band flux comes from an unresolved point source with a size ≤ 35 pc, and the rest is extended emission from the narrow line region.

Here we report new mid-infrared interferometric observations of NGC 4151 which clearly resolve a thermal structure.

4.2. Instrument, observations and data reduction

The observation procedure was as described in Section 2.5

Observations were taken in the nights of April 21 and 23, 2008 with $BL = 61$ m and 89 m at position angles of $103°$ and $81°$ respectively. These provide effective spatial resolutions[4] ($\lambda/3BL$) at $10.3\mu m$ of 11 and 7 milliarcseconds (mas), respectively. The calibrators HD 133582 and HD 94336 were selected to be very close in airmass with $\Delta z \lesssim 0.15$. This is especially important for NGC 4151 (DEC $\approx +40°$) which, at Paranal, never rises higher than $\approx 25°$ ($z \gtrsim 2.3$) above the horizon. The northern declination of the source also limits the projected baselines and fringe patterns to essentially East-West orientation at Paranal (see Figure 4.2). The N band spectrum of HD 133582 (K2III) was taken to follow a Rayleigh-Jeans law, while that of HD 94336 (MIII) was taken from Cohen et al. (1999).

Data reduction was performed with MIA+EWS as described in Section 2.6[5].

---

[4]see Section 2.2.5
[5]For the data reduction of this source, the nightly build from Dec 02 2009 was used.

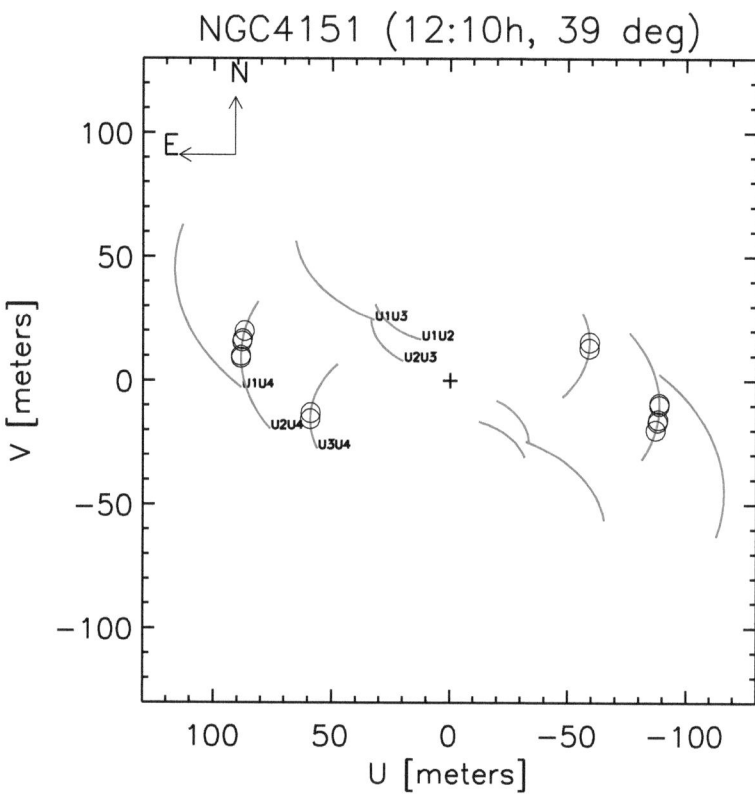

Figure 4.2.: Observed (u, v) coverage for NGC 4151. The open circles denote the (u, v) coordinates of the fringe tracks on the two baselines. Their diameter corresponds to the telescope diameter D of the VLT UTs, $D = 8$m. The thin lines are the (u, v) tracks that are traced by the various telescope combinations as a result of earth's motion. They are plotted for elevations $> 20°$. (u, v) plane tracks are followed CCW if $\delta < 0$, CW if $\delta > 0$.

4.3. Results and modeling

4.3.1. Single-dish spectrum

The resulting single-dish spectrum with an effective aperture of ≈ 300 mas is shown in Figure 4.3 together with a Spitzer IRS spectrum for this source (observed 8 Apr 2004; aperture $\approx 3.5''$, Weedman et al. 2005). The higher Spitzer flux most likely indicates emission from the Narrow Line Region (cf. Radomski et al. 2003).

The color temperature of the spectrum is (285^{+25}_{-50}) K, slightly warmer than the 201 K estimated from IRAS 12 and 25 μm data by Soifer et al. (2003). The reason for this is probably that the large aperture IRAS data contain not only the warm nuclear material but also colder material from larger scales. Additionally, we detect the [Ne II] 12.81 μm line commonly seen in star forming regions. The [S IV] 10.51 μm line, clearly seen in the Spitzer spectrum, is not significant in the MIDI spectrum. The errors in the MIDI spectrum arise from incomplete thermal background subtraction and hence rise steeply with increasing wavelength.

4.3.2. Correlated spectra

The two correlated flux spectra are shown in Figure 4.4.

The correlated flux observed with $BL = 61$m rises from (0.13 ± 0.01) Jy at 8.5 μm to (0.43 ± 0.05) Jy at 12.5 μm. The second spectrum $(BL = 89$m$)$, rises from (0.12 ± 0.02) Jy at 8.5 μm to (0.30 ± 0.05) Jy at 12.5 μm. Correlated flux uncertainties arise primarily from background photon noise and increase with wavelength but are smaller than the single dish errors.

The correlated flux on the shorter baseline (the one that has a higher flux) shows a broad "bump" between 9 μm and 12 μm that we interpret as a silicate emission feature (see section 4.3.3). No [Ne II] emission is seen, indicating that this arises on a scale that is fully resolved by the interferometer (i.e. $\gtrsim 20$ mas ≈ 1.3 pc). The fact that the correlated flux is lower on the longer baseline is a clear sign that the source is resolved by the interferometer.

Since the resolution of the interferometer $\theta_{min} \approx \lambda/3BL$ changes with wavelength, the correlated flux reflects both the source spectrum and the source structure. It is not possible to draw any conclusions from the spectral slope of a correlated flux spectrum without assuming a source geometry.

4.3.3. A possible silicate emission feature

Although the silicate absorption feature has often been detected in type 2 nuclei, the emission feature, predicted for type 1 nuclei by torus models (e.g. Pier & Krolik 1992; Schartmann et al. 2005), has not been detected except in a handful of objects, most of them quasars (Hao et al. 2005; Weedman et al. 2005; Buchanan et al. 2006).

As noted by Weedman et al. (2005) and Buchanan et al. (2006), Spitzer spectra (with an aperture of $\approx 3.5''$) of NGC 4151 show weak excess emission at 10 μm and 18 μm that

4. NGC 4151: The first resolved nuclear dust in a type 1 AGN

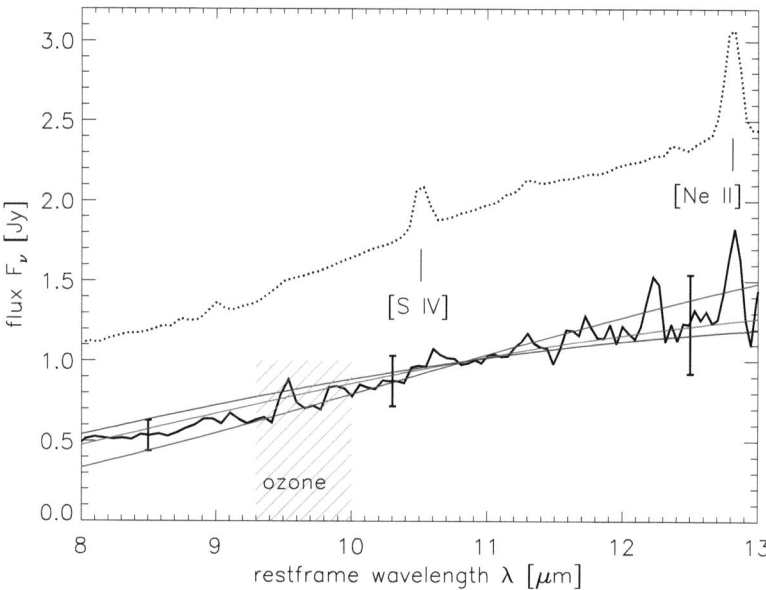

Figure 4.3.: Single-dish spectra for NGC 4151. Spitzer spectrum (aperture: 3.5″, dotted, Weedman et al. 2005), MIDI spectrum (aperture: 0.3″, black line with error bars). The MIDI single-dish errors were taken as the error of the mean from five observations. Also plotted are blackbody emission curves at $T = 235, 285, 310$ K (green, red, blue). The region of atmospheric ozone absorption, between 9.3 and 10 μm (hatched), is uncertain and not taken into account for the later analysis.

4.3. Results and modeling

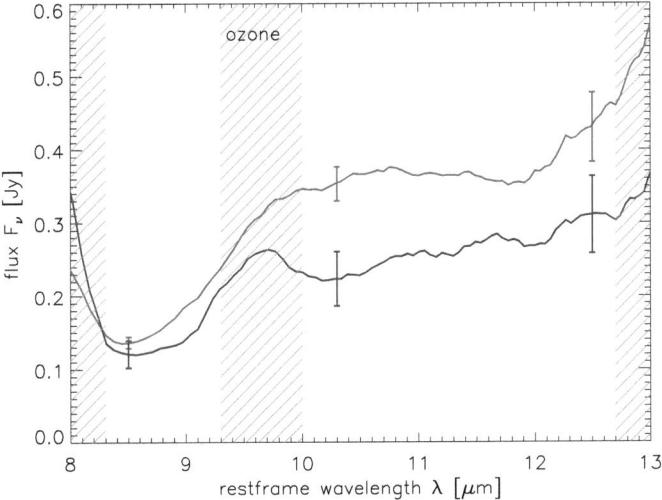

Figure 4.4.: Smoothed ($\Delta\lambda \approx 0.3\mu m$) correlated MIDI spectra at two different East-West projected baselines: 61m (red, average of two fringe track observations) and 89m (blue, average of three fringe track observations). The errors are the errors of the mean of the individual observations. The region of atmospheric ozone absorption, between 9.3 and 10 μm (hatched), is uncertain and not taken into account for the later analysis, as are the regions $\lambda < 8.3\mu m$ and $\lambda > 12.7\mu m$ (hatched), which have very low SNR.

is most easily seen when plotted as $\nu^2 F_\nu$ with a sufficiently large wavelength range. In Figure 4.5 we plotted our spectra (that are limited to the atmospheric N band) in such a way together with a Spitzer spectrum.

The hypothesis of silicate emission is not inconsistent with our single dish spectrum (aperture $\approx 0.3''$) but since this spectrum suffers from incomplete background subtraction, especially at longer wavelengths, it is probably hidden in the resulting uncertainties. The emission feature seems to be most prominent in our observations on the 61m baseline observation and is clearly not detected on the 89m baseline.

This suggests that at least some fraction of the silicate emission feature observed in type 1 AGNs is located on scales ≈ 1 pc as derived from the 10.5 μm Gaussian model (see 4.3.4). Recent observation of bright quasars, in combination with dusty Narrow Line Region models, have suggested that the silicate emission feature may be produced in the Narrow Line Region, although a contribution from the parsec-scale is not excluded (Schweitzer et al. 2008).

4.3.4. Simple Gaussian model

With the limited baselines available from Paranal, it is not possible to reconstruct an *image* of NGC 4151 from our data. Instead we consider simple model distributions of the emission on the sky and compare the predicted interferometric and single dish spectra with our measurements in order to fix parameters in such a model (see Section 2.2.6). We chose a model containing an unresolved point source (flux F_p) and an extended Gaussian distribution (flux F_g, FWHM θ). Although we might expect the mid-IR brightness distribution in Sy 1 galaxies to have a hole in the middle, the fluxes do not change as long as the hole, whose radius is determined by the sublimation radius of the dust, is unresolved. This is certainly the case in NGC 4151 where $r_{\rm rev} = 0.04$ pc, roughly four times smaller than our highest resolution observation (see section 4.4). An upper limit to the size of our point source is given by the effective resolution of the interferometer; this corresponds to a diameter of $\approx \lambda/3BL \approx 7$ mas, i.e. a radius of ≈ 0.2 pc, at 10.3 μm.

Because of the East-West baseline orientation, the North-South distribution of emission is undetermined. Equivalently, we assume the source to be circularly symmetric on the sky.

The model correlated flux density at wavelength λ is then given by

$$F_\nu(\lambda) = F_p(\lambda) + F_g(\lambda) \cdot \exp(-(BL/\lambda \cdot \pi/2 \cdot \theta)^2/\ln(2)). \qquad (4.1)$$

With the given data points this model is uniquely determined. We calculated the parameters of such a model separately at 8.5, 10.3 and 12.5 μm where we are safely away from the regions of very low signal to noise (at the edges of the N band) and the ozone feature. At 10.3 μm the parameter values may be affected by the silicate feature. The modeled visibilities are shown in Figure 4.6 and the resulting parameters are given in Table 4.1.

The errors of these parameters were estimated from a Monte-Carlo simulation in which we resampled our data by randomly placing 10 000 measurements in a Gaussian distri-

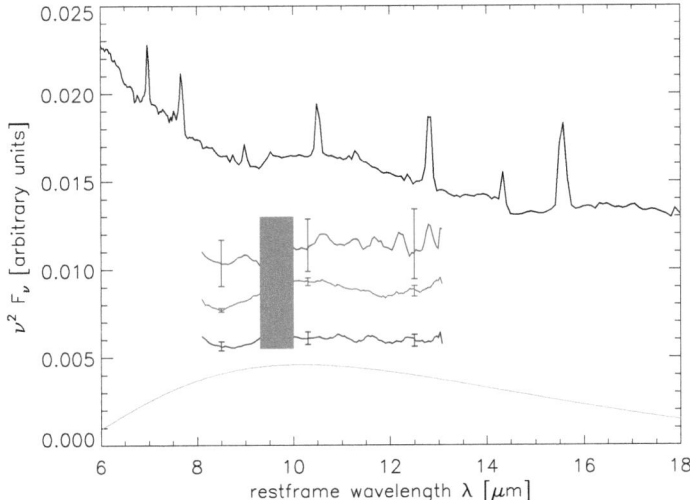

Figure 4.5.: The silicate emission feature as seen in $\nu^2 F_\nu$. In the Spitzer spectrum (black, Weedman et al. 2005), a broad weak emission feature at (11 ± 1.5) μm is clearly seen. The three MIDI spectra (green: single-dish, red: 61m baseline, blue: 89m baseline) are plotted with offsets. The grey line shows a 285 K blackbody. The region between 9.3μm and 10 μm is hard to calibrate in the MIDI spectra due to the atmospheric ozone feature and has been greyed to not mislead the eye.

4. NGC 4151: The first resolved nuclear dust in a type 1 AGN

Table 4.1.: Parameters for the Gaussian models (see text).

λ [μm]	F_g [Jy]	θ [mas] (diameter [pc])	F_p [Jy]
8.5 ± 0.2	0.41 ± 0.10	$29^{+3}{}_{-6}$ ($2.0^{+0.2}{}_{-0.4}$)	0.119 ± 0.016
10.3 ± 0.25	0.70 ± 0.16	23 ± 4 (1.5 ± 0.3)	0.194 ± 0.061
12.5 ± 0.3	1.03 ± 0.30	32 ± 6 (2.1 ± 0.4)	0.290 ± 0.070

bution around the measured value with the σ as determined from our data reduction. The parameter errors are then given by the standard deviations of the resulting model solutions to the simulated data.

4.4. Discussion

4.4.1. The extended source and the Sy 1 / Sy 2 paradigm

In the strictest version of unified models, we expect for both a Sy 1 and a Sy 2 galaxy an extended dust structure with the same size, morphology and temperature distribution (at a given UV luminosity L_{UV}). In less strict versions this is only true statistically (e.g. Elitzur & Shlosman 2006). Additionally, Sy 1 galaxies should have an unobscured point source (the unresolved accretion disk and inner rim of the torus) – but the relative strength of these two components in the mid-IR is model-dependent.

To test this, we can compare our observations with other MIDI observations: In the Circinus galaxy ($L_{UV} \approx 4 \times 10^{36}$ W, $L_{torus} \approx 5 \times 10^{35}$ W), Tristram et al. (2007) found a warm ($T \approx 330$ K) disk with major axis FWHM ≈ 0.4 pc and a larger, similarly warm, component of ≈ 2 pc FWHM. In NGC 1068 ($L_{UV} \approx 3 \times 10^{37}$ W, $L_{torus} \approx 10^{37}$ W), Raban et al. (2009) found a hot ($T \approx 800$ K) disk of 1.35 x 0.45 pc and a warm component 3 × 4 pc in FWHM. They identified the disks with the densest parts of the torus of the unified model.

For NGC 4151 ($L_{UV} \approx 1.5 \times 10^{36}$ W – variable (from NED), $L_{torus} \approx 4\pi D^2 \nu F_g \approx 6 \times 10^{35}$ W) we determined a torus size (FWHM) of $\approx (2.0 \pm 0.4)$ pc and a dust temperature of (285^{+25}_{-50}) K.

When scaled to the accretion disk luminosity L_{UV}, these values agree well with the torus sizes ($r \approx L_{UV}^{1/2}$) and torus luminosities ($L_{torus} \approx L_{UV}$) of the two Seyfert 2 galaxies and the temperatures are also very similar. Note, however, that the torus luminosity used here is calculated from the 12μm flux density, not taking into account emission at longer wavelengths.

4.4.2. Greybody models and the nature of the extended source

In addition to the single-wavelength Gaussian models discussed above, we can also connect the various wavelengths together by constructing *greybody* models where we assume a

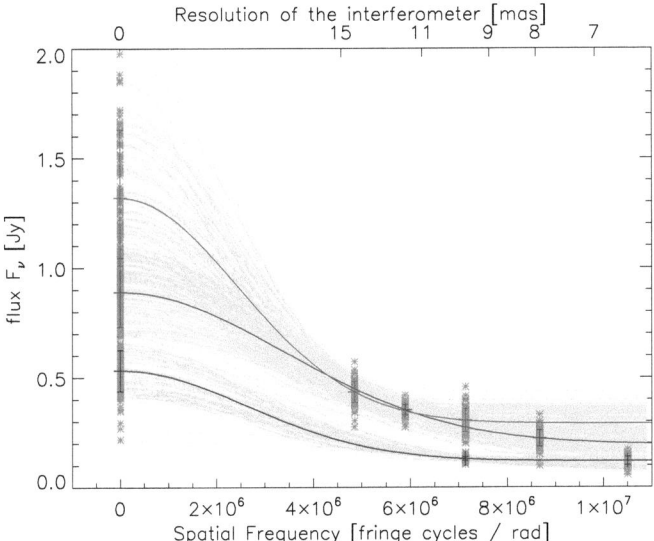

Figure 4.6.: Gaussian model for NGC 4151. On the upper x axis the resolution (i.e. sensitivity to model parameters) of the interferometer $\lambda/3BL$ is given. The three curves correspond to uniquely determined models at 8.5 μm (blue), 10.3 μm (green) and 12.5 μm (red). Data points and errors are taken from the single-dish (i.e. "0-baseline", Figure 4.3) and correlated fluxes (the other two data points per wavelength, Figure 4.4). The many thin lines represent the Monte-Carlo-like error determination for the Gauss model. For details see text.

4. NGC 4151: The first resolved nuclear dust in a type 1 AGN

smooth power-law dependence of temperature with radius: $T(r) = T_0(r/R_{\min})^{-\alpha}$, and an essentially constant emissivity ϵ with radius and wavelength. The model flux is then given by

$$F_\nu = \int I_\nu \, d\Omega = \frac{2\pi}{D^2}\int I_\nu r \, dr = \frac{2\pi}{D^2}\int_{R_{\min}}^{R_{\max}} \frac{2\pi h \nu^3}{c^2}\frac{\epsilon}{\exp\left(\frac{h\nu}{kT(r)}\right)-1} dr \qquad (4.2)$$

where the inner integration boundary $R_{\min}$ is given by the reverberation radius ≈ 0.04 pc and the outer boundary is set to the resolved scale $R_{\max} \approx 0.2$ pc / 0.35 pc (for the correlated flux at 8.5 μm and 12.5 μm respectively) and 1 pc (for the total flux). The integral is very sensitive to the inner boundary (where the temperature is maximal), the outer boundary is not so critical. The temperature at $R_{\min}$, T_0, is fixed to the value derived by Riffel et al. (2009) from a fit to the near-IR spectrum (see below).

With such models we get acceptable fits (reduced $\chi^2 \approx 1$) for both the total and the correlated flux spectra with $T(1\text{pc}) \approx 250$ K and $\alpha \approx 1/2.8 \approx 0.36$. This value of α is consistent with dust receiving direct radiation from a central source (e.g. Barvainis 1987), i.e. an optically thin medium with optically thick clumps in it.

These models can be extrapolated to shorter wavelengths to check their consistency with the K band measurements. All plausible extrapolations of the MIDI data yield K band fluxes of < 10 mJy, much less than observed by Swain et al. (2003) and Riffel et al. (2009) (see below). On the basis of these greybody models, we therefore conclude that the K band emission arises from structures which can probably not be extrapolated from the larger structure seen in the N band.

From the greybody models one can further infer an emissivity $\epsilon \approx 10^{-1}$ – similar to what has been seen in NGC 1068 and Circinus.

To conclude: The resolved nuclear mid-IR structure in NGC 4151 has a size, temperature and emissivity that is comparable to those in type 2 objects where the existence of clumpy tori is established. Apart from the similarities with type 2 tori and the temperature profile of the greybody models, there is further evidence for clumpiness in the NGC 4151 torus from radio observations: Mundell et al. (2003) measured HI absorption against the radio jet ($PA \approx 77°$) and found a structure of ≈ 3 pc in size. From the velocities they further suggest that the gas is distributed in clumps. Warm dust possibly traces the HI gas in the mid-IR.

It therefore seems reasonable to identify the warm dust structure resolved now in NGC 4151 with the clumpy tori seen in Sy 2 galaxies. Due to the limited observing geometry and the limited amount of observations we cannot reconstruct its apparent shape nor can we constrain a model with more than the two components discussed above.

4.4.3. The point source and its relation to K band measurements

K band interferometry measurements by Swain et al. (2003) revealed a marginally resolved source, compatible with hot dust at the sublimation radius of ≈ 0.05 pc. This was confirmed in more recent observations with higher significance (Pott et al. 2010). Reverberation measurements by Minezaki et al. (2004) and Koshida et al. (2009) find lag times Δt between the UV/optical continuum and the K band corresponding to a radius of ≈ 0.04 pc (variable) which they interpret as the sublimation radius of dust.

Riffel et al. (2009) measured a flux of ≈ 65 mJy at 2.2 μm and from spectral fitting they found that, at 2.2 μm, this flux is entirely dominated by a blackbody with a temperature of (1285 ± 50) K – again consistent with hot dust at the sublimation radius and the K band interferometry, taking into account that $L_{\rm UV}$ is variable by at least a factor of 10 (Ulrich 2000). To account for that variability when comparing our observations with these K band observations of different epochs, we looked at the X-Ray flux[6] as a proxy for the UV–optical radiation and find that, at the date of our observations, the source was probably in a higher state than at the time of Riffel et al. (2009)'s observations and emitted ≈ 0.2 Jy in the K band (see Figure 4.7).

Since the flux density F_ν of a ≈ 1285 K blackbody is roughly the same at 2.2μm as at 8.5 μm, we can compare the flux in the K band (≈ 0.2 Jy at the time of our measurement) with our point source flux at 8.5 μm ($\approx (0.119 \pm 0.016)$ Jy). From this it seems likely that hot dust is contributing to our point source at 8.5 μm. The spectrum of the point source rises by more than a factor of two from 8.5 μm to 12.5 μm, however, and thus requires emission from an additional small, "red" component. This could be core synchrotron emission, although such sources usually have flat spectra, or emission from a small, cool, optically thick central dust structure, possibly shadowed from direct accretion disk radiation.

4.5. Conclusions

Using mid-IR interferometry, we resolve a warm dusty structure in NGC 4151. Its FWHM size (2.0 ± 0.4 pc – from comparing the data with a Gaussian model), temperature (285^{+25}_{-50} K) and emissivity (≈ 0.1) are in good agreement with the clumpy tori seen in type 2 AGNs and are thus consistent with the unified model of Active Galaxies. Excess emission around 10.5 μm on the intermediate baseline indicates that silicate emission might arise from scales of ≈ 1 pc in AGNs.

Using simple models we compare our mid-IR fluxes with observations in the K band and find that the structure we resolve is probably not the smooth continuation of the nuclear source detected in the K band

Due to the limited number of measurements, no two-dimensional information can be gathered and questions about the unified model (such as: is the dust structure in Sy 1 galaxies thick and torus-like or rather disk-shaped?) remain unanswered. Since nuclear dust distributions are different even within the same class of AGNs, the ultimate question whether or not the unified model is valid will not be answered before a statistical study of numerous tori is performed. Such a study is presented in Chapter 5.

---

[6]Data from the All-Sky-Monitor (ASM) on the Rossi X-Ray Timing Explorer (RXTE), available online.

4. NGC 4151: The first resolved nuclear dust in a type 1 AGN

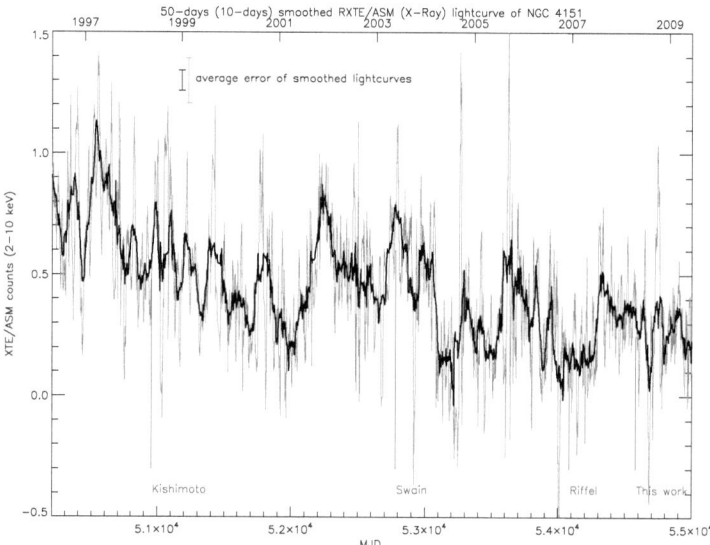

Figure 4.7.: X-Ray lightcurve from the All-Sky-Monitor (ASM) onboard the Rossi X-Ray Timing Explorer (RXTE), summed over all bands of the instrument and smoothed by 50 days (black curve) and 10 days (grey curve), respectively. The lower x axis displays the Modified Julian Date (MJD). The nights of relevant observations are marked: Riffel et al. (2009) seem to have observed the source in a rather low state compared to the earlier observations of Kishimoto et al. (2007) and Swain et al. (2003). See Figure 3.14 for the X-Ray-IR-variability correlation in Centaurus A.

5. The MIDI Large Programme: A statistical sample of resolved AGN tori

5.1. Introduction[1]

The first detailed interferometric studies of the brightest AGNs in the mid-IR have resolved their nuclear dust distributions (Circinus galaxy, Tristram et al. (2007) and NGC 1068, Raban et al. (2009)). They showed that warm dust exists on the parsec scale, that their geometry and temperature are not universal and that their structure is probably clumpy (a summary of these studies has been given in Section 1.4).

In addition to these two mid-IR brightest sources, another dozen, weaker, galaxies had been examined in the MIDI GTO time, but generally only one or two (u, v)-points were observed (Tristram et al. 2009). Therefore, this 'snapshot' survey served more to prove the observability of weaker targets than to provide a statistical basis for size and structure analysis.

It was clear that a large and systematic observational campaign was needed to collect the basic observational information necessary to understand dusty tori on a statistical basis. This is the aim of the MIDI AGN Large Programme (LP).[2]

The questions to be addressed in the LP are:

1. How does the measured mid-IR morphology of AGN tori depend on wavelength, nuclear orientation and luminosity? This is the key to understanding the radiation transfer effects in the dust structure. Is there a common torus size – AGN luminosity relation for all types of AGNs as suggested by the GTO study (Tristram et al. 2009)? Understanding and properly calibrating this relation with local AGNs is the only way to safely apply the relation to the wealth of distant spatially unresolved AGNs.

2. Are other "parameters" important for the morphology, such as mass feeding rate from circumstellar star clusters, or dust chemistry?

3. Is the Sy 1 and Sy 2 dichotomy only an orientation effect or a simplified view of a wide variety of different intrinsic morphologies? Are 10 μm silicate features always seen in absorption in Sy 2s and in emission in Sy 1s?

---

[1] This section and subsection 5.2.1 were adapted from the Large Programme proposal 184.B-0832 (PI: Meisenheimer)

[2] So far this is the only Large Programme for any VLTI instrument.

4. Is the apparent "two component" structure of a compact dense disk and an extended almost round (spherical?) distribution (see Figure 1.4, left panel) a general property? Is the dust distribution patchy / clumpy (as found in Circinus)? In which AGNs does maser emission coincide with the dust emission?

5. Is the inner rim at the sublimation radius? So far hot dust ($T \sim 900\,\mathrm{K}$) has been detected in only one object (NGC 1068).

Some of these questions (e.g. on the size–luminosity relation) can be answered directly from the interferometric data, others, such as questions on the structural properties of the dust (e.g. clump sizes and distributions), require a more detailed understanding of the spectra through radiative transfer models. The most far-reaching astrophysical questions, involving accretion mechanisms from the kpc scale of the nuclear star cluster to the parsec-scale dust and further in, require comparisons with hydrodynamical models, or at least a study of the physical mechanisms responsible on these scales. The various participating researchers in the Large Programme have planned to perform these studies in the near future.

Here, the first full data reduction of the Large Programme, a set of one-dimensional geometrical models and some direct implications from these model fits are presented.

5.2. Observations and Data Reduction

5.2.1. Observational strategy

While the size and shape of the nuclear dust distribution can be derived from very few (u, v) points, a more detailed study is required to investigate the *structure* of the dust distribution and particularly measure clumpiness and warping.

To answer both the questions on detailed structures and on statistical relationships within reasonable time limits, the Large Programme therefore had two objectives:

1. Determine the basic properties for a sample of 8 sources (the *extended snapshot* subsample) and

2. obtain *detailed maps* for three more sources where preliminary data indicated that MIDI achieves good spatial resolution.

The three galaxies for which detailed maps already existed (NGC 1068, the Circinus galaxy and Centaurus A) are some of the closest Active Galaxies. In order to extend the sample, more distant galaxies with higher luminosities needed to be included. While, on first sight, it may appear that the ability to resolve the dust structure is a strong function of distance, this is not the case, if the "torus" size s follows the very basic scaling relation expected for centrally heated dust with heating luminosity L, $s \propto L^{1/2}$: Then, the apparent size at distance D is $\Theta = s/D = \mathrm{const} \cdot L^{1/2}/D$. The luminosity L relates to the observed flux F as $L = 4\pi D^2 F$. Thus $\Theta \propto F^{1/2}$, independent of the distance D.

Studies of more distant AGNs therefore are feasible with the VLTI, if the above relations hold.

5.2.2. Target list

Some of the LP targets had been observed before in the GTO programmes and were included to complement their (u, v) coverage. Others had not been observed with MIDI before. For selecting those sources, similar criteria had been applied as for the original GTO study: Besides being well observable from Paranal (Declination $\lesssim 25°$), the sources were required to have an unresolved core flux $\gtrsim 200$ mJy[3] at ≈ 12 μm as determined from high-resolution imaging observations (such as Krabbe et al. 2001; Siebenmorgen et al. 2004; Gorjian et al. 2004; Galliano et al. 2005; Haas et al. 2007; Raban et al. 2008; Horst et al. 2009; Gandhi et al. 2009; Prieto et al. 2010; Reunanen et al. 2010). To predict observability with MIDI, it is essential to take nuclear fluxes only from high-resolution observations, as large aperture observations (such as from the IRAS, ISO or Spitzer satellites) are very often contaminated by non-nuclear mid-infrared emission, e.g. from starburst regions.

These selection criteria resulted in a very high success rate of MIDI observability (Raban et al. 2008; Tristram et al. 2009).

In the LP, one quasar, six type 1 and seven type 2 galaxies have been observed[4]. Most of them have total fluxes $\lesssim 1$ Jy and all of them have correlated fluxes significantly < 1 Jy. They are therefore among the weakest sources ever observed with MIDI. The (angular size) distances of these sources range from 18 Mpc (NGC 1365) to the cosmological distance of 546 Mpc (3C 273, $z = 0.158$) with the median at 52 Mpc. The most relevant information of the Large Programme targets has been collected in Table 5.1 and the individual targets are introduced in Section 5.4.1.

5.2.3. The observations

Observations were carried out in Visitor Mode between December 2009 and August 2010. Since the targets were much fainter than supported by standard ESO service mode observations, Visitor Mode was required in order to ensure an optimum observing strategy (see Section 2.5). For the following analysis all other data for the LP targets, that was available through the ESO archive, has also been taken into account. The observation logs together with the ESO programme numbers are given in Burtscher (2011).

A total of 13.1 nights, corresponding to 151.5 hours (including twilight time[5]), were scheduled of which about 25 % were lost due to technical problems and bad weather. The

---
[3]The GTO criterium was $\gtrsim 400$ mJy.
[4]NGC 5506, NGC 5995 and 3C 273 were not part of the original target list, however. NGC 5506 has been included as a replacement target for Mrk 463 E that proved to be unobservable; NGC 5995 and 3C 273 have been added as backup targets.
[5]ESO's conversion factor between nights and hours only counts hours after/before astronomical twilight, but mid-IR observations can already be started, respectively carried on, in twilight time.

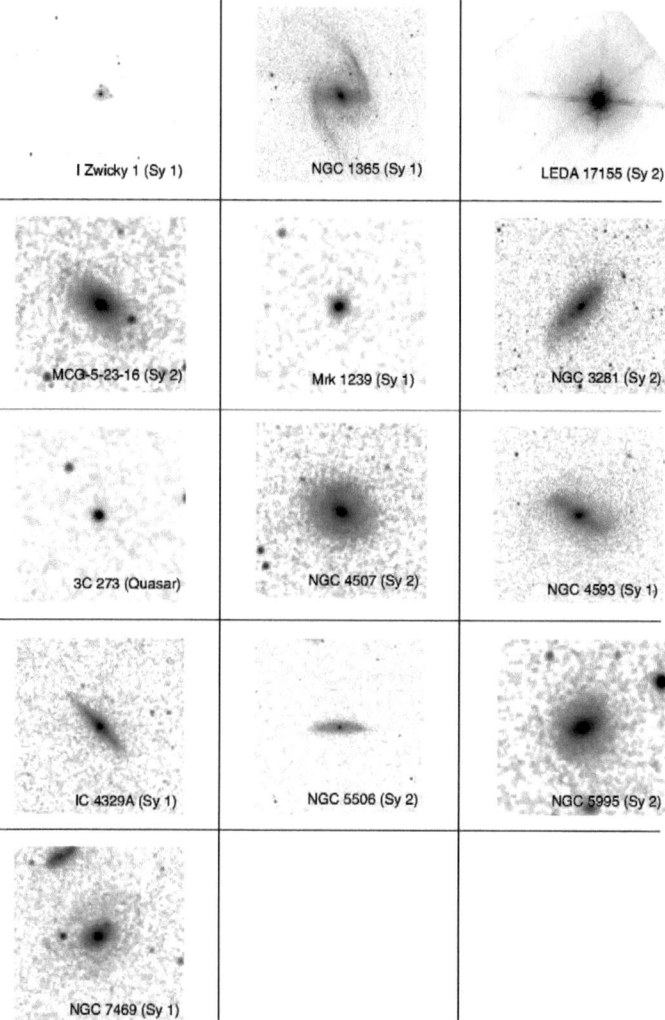

Figure 5.1.: Inverted color images of the Large Programme targets at nearly identical resolutions and wavelengths (with the exception of I Zwicky 1 that is a BVR composite from the EFOSC2/ESO 3.6 m telescope, Courtesy I. Saviane and the EFOSC2 gallery and LEDA 17155 (IRAS 05189-2524) that is an HST NICMOS image (courtesy N. Scoville) from Scoville et al. (2000)). All others are 2MASS JHK_s composite images (Credit: NASA/IPAC Infrared Science Archive).

The box sizes of the images are 5' (I Zwicky 1), 9.6' (NGC 1365), 12' (LEDA 17155), 1.8' (MCG-5-23-16), 1.7' (Mrk 1239), 5' (NGC 3281), 1.5' (3C 273), 2.5' (NGC 4507), 5.1' (NGC 4593), 3' (IC 4329A), 6.5' (NGC 5506), 1.5' (NGC 5995), 2.6' (NGC 7469)

Table 5.1.: Large Programme targets and other AGNs studied with MIDI: NGC 1068 (Raban et al. 2009), NGC 4151 (Chapter 4) and the Circinus galaxy (Tristram 2007). Explanations of the columns are given below the table.

Name	RA [h m s]	DEC [° ′ ″]	type (1)	type (2)	D [Mpc]	scale [mas/pc]	$F_\nu^{12.5\mu m}$ [Jy]	Coudé star Ψ[″]	Coudé star V [mag]	$n_{(u,v)}$	$N_{(u,v)}$
I Zw 1	00 53 35.1	+12 41 34	Sy 1	Sy 1	222	0.93	0.44±0.04	0	14.1	9	18
NGC 1365	03 33 36.4	-36 08 25	Sy 1.8	Sy 1	18.1 ± 2.6	11.40	0.41±0.04	0	13.5	12	25
LEDA 17155	05 21 01.7	-25 21 45	Sy 1h	Sy 2	167	1.24	0.31±0.06	0	16.5	6	8
MCG-05-23-16	09 47 40.2	-30 56 54	Sy 1i	Sy 2	38.8	5.32	0.64±0.05	11	13.5	11	22
Mrk 1239	09 52 19.1	-01 36 43	NL Sy 1	Sy 1	84.5	2.44	0.43±0.05	0	14.4	5	9
NGC 3281	10 31 52.1	-34 51 13	Sy 2	Sy 2	47.6	4.33	0.63±0.09	15	15.7	3	7
3C 273	12 29 06.7	+02 03 08	Sy 1.0	Quasar	546	0.38	0.33±0.09	0	12.9	3	4
NGC 4507	12 35 36.6	-39 54 33	Sy 1h	Sy 2	51.7	3.99	0.39±0.08	28	16.0	4	7
NGC 4593	12 39 39.4	-05 20 39	Sy 1.0	Sy 1	44.0 ± 6.4	4.69	0.20±0.07	0	13.9	3	6
IC 4329 A	13 49 19.3	-30 18 34	Sy 1.2	Sy 1	68.3	3.02	1.00±0.04	0	12.9	8	12
Mrk 463 E	13 56 02.9	+18 22 19	Sy 1h	Sy 2	197	1.05	0.34	0	14.3	0	0
NGC 5506	14 13 15.0	-03 12 27	Sy 1i	Sy 2	28.9 ± 0.2	7.14	1.16±0.08	0	12.4	4	8
NGC 5995	15 48 25.0	-13 45 28	Sy 1.9	Sy 2	102	2.02	0.35±0.13	26	13.0	2	4
NGC 7469	23 03 15.6	+08 52 26	Sy 1	Sy 1	60.9	3.39	0.44±0.08	0	13.3	2	8
NGC 1068	02 42 40.7	-00 00 48	Sy 1h	Sy 2	14.4	14.32	16.00±1.00	0	12.0	12	16
NGC 4151	12 10 32.6	+39 24 21	Sy 1.5	Sy 1	14	14.73	1.30±0.30	0	12.3	3	7
Circinus	14 13 09.9	-65 20 21	Sy 1h	Sy 2	4	51.57	14.00±0.50	50	12.5	11	21

(1) AGN type as classified in NED (Véron-Cetty & Véron 2006): The classification of intermediate Seyfert types ($1 <$ type < 2) is defined by the ratio of Hβ to [OIII]λ5007 fluxes. Sy 1h are 'hidden' Seyfert 1 galaxies (i.e. where broad lines have been detected in polarized light), In the Sy 1i type, broad Pa β lines are detected
(2) AGN type from SIMBAD
D (without given uncertainty): angular-size distance derived from redshift using the CMB reference frame and a concordance ΛCDM cosmology, D (with given uncertainty): redshift-independent distance measurement (average over several observations, see NED), the distances of NGC 1068, NGC 4151 and of the Circinus galaxy have been taken from the respective publications.
$F_\nu^{12.5\mu m}$: flux at $12.5 \pm 0.2\mu m$ as seen by MIDI (the Mrk 463E flux is from Raban et al. (2008))
Ψ: Coudé (MACAO) guide star separation from science target (0: guiding on nucleus)
$n_{(u,v)}$: number of good fringe tracks at (u, v) points separated by $\gtrsim 1$ telescope diameter ($D = 8$ m) from each other
$N_{(u,v)}$: total number of good fringe tracks

5. The MIDI Large Programme: A statistical sample of resolved AGN tori

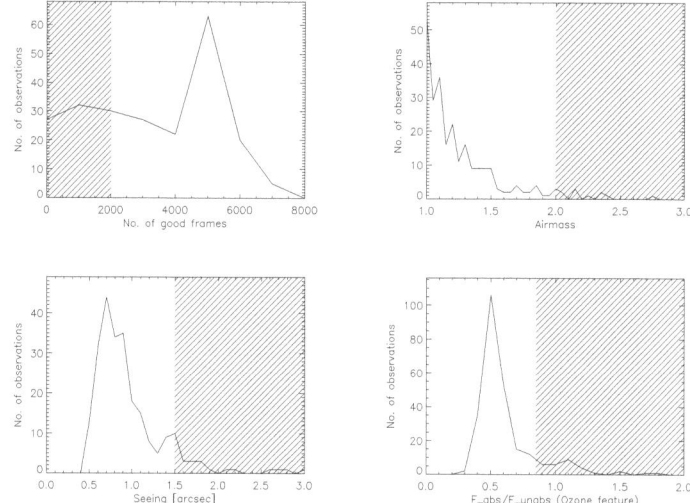

Figure 5.2.: Histogram for the quality of the fringe track observations measured by the number of good frames (top left), airmass (top right), seeing (bottom left) and $F_{\mathrm{abs}}/F_{\mathrm{unabs}}$ in the telluric Ozone feature (bottom right). Observations in the hatched areas were rejected. See text for details.

programme profited from two excellent low-seeing and large-coherent-time nights (2010-03-26 and 2010-03-27), experienced many average nights (with less than 1 hour losses due to technical problems or weather) and a relatively large number of very bad nights (such as most of the nights in May 2010) during which strong wind, extreme seeing, fog and even rain made observations impossible.

5.2.4. Data reduction, selection and handling

Data selection Selection criteria were applied to both raw and reduced data to reject bad observations and spurious results. The criteria were based on experience with other datasets and tests of systematic errors (see Section 2.7.4) and they were rather strict in order to get a *first reliable overview* of the whole LP dataset, not to keep the maximum number of datasets.

Seven fringe tracks were rejected on the raw-data level because the telescope was in chopping mode (such data cannot be reduced reliably) and two were rejected because they were accidentally taken in the wrong tracking mode[6] and no suitable calibrators were available.

For the data selection in the reduced data stage, automatic selection criteria were applied to prevent (or at least control) the introduction of biases due to the selection. Out of the 253 fringe tracks (228 from the Large Programme and 25 from the archive) taken into account, 138 were identified as good and 115 were flagged as bad (Figure 5.2) for the following reasons (some tracks were rejected for multiple reasons), see the observation logs in Burtscher (2011):

- number of good frames (as determined by EWS, see Section 2.6) N_{good} was required to be > 2000. For the weak sources ($F_{\text{corr}} \approx 300$ mJy), 8000 frames roughly correspond to a SNR of only 5. If there are less than 2000 frames, not only does the SNR decrease to unacceptable levels, but often this is also indicative of other, maybe undetected, problems with this dataset. (59 tracks were rejected for this reason)

- The airmass z was required to be < 2.0. For very high airmasses (and weak sources), the adaptive optics correction by MACAO works considerably less well, possibly loosing overlap between the two beams. In some cases, the correlated fluxes have reached unrealistically low values for very high airmass observations. (13 tracks)

- Only observations taken at a seeing $< 1.5''$ were accepted. The seeing values were taken from the DIMM observations in the optical near zenith. While MIDI observes in a different waveband and possibly at a different location on the sky than the DIMM, this value proved to correlate with the calibration errors for MIDI (see Section 2.7.4). Besides affecting the beam overlap, large seeing values also imply low values for the atmospheric coherence time, leading to correlation losses. (40 tracks)

- The Ozone feature depth was required to be well visible above the noise in the raw counts: The telluric Ozone feature should produce a pronounced decrease of flux at (9.65 ± 0.25) μm so that the ratio between the measured flux, F_{abs}, to the expected unabsorbed flux, $F_{\text{unabs}}^{\text{expected}}$, should be ≈ 0.4. By linearly interpolating between (9.0 ± 0.1) μm and (10.3 ± 0.1) μm, a rough estimate of the "unabsorbed" count rate, F_{unabs} is found. If $F_{\text{abs}}/F_{\text{unabs}} > 0.85$, not a large portion of the signal went through the earth's atmosphere and the observation is probably spurious. (37 tracks)

Apart from these automatic selection criteria, data were manually inspected if the calibrated fluxes showed unexpected results. This way 13 fringe track observations were manually rejected due to clouds (as seen from the ESO ambient conditions database), obvious signs of correlation losses (extremely low fluxes at 8 μm, not seen in other correlated flux spectra of this source), extremely low SNRs or large variations in the OPD during the fringe track.

---

[6] Weak source fringe tracks must be taken in off-zero tracking mode to maximize SNR, see Chapter 2.

5. The MIDI Large Programme: A statistical sample of resolved AGN tori

For the single-dish spectra, similar criteria were applied, but the seeing limit was relaxed to 2.0. Out of 148 single-dish spectra (130 from the Large Programme, 18 from the archive), 79 were identified as good and 69 as bad. Out of the latter, 37 were flagged for their excessive flux in the Ozone feature, 5 due to high seeing, 6 due to high airmass. Further 24 single-dish spectra were rejected due to very uneven backgrounds, clouds or problems that affected only one of the two photometric channels.

The bad data flags are included in the observation logs (Burtscher 2011).

Data reduction The data reduction procedure followed the outline given in Section 2.6. For the very weak targets observed during the Large Programme, the most recent (2011) versions of MIA+EWS produced different results than older versions (especially the ones released before 2009) because of the various modifications of the data reduction routines for weak targets as described in Section 2.6.

Data reduction and handling Data reduction of such a large amount of data (ca. 300 GB of raw data or roughly 10 million individual frames) requires a considerable amount of book-keeping. For this purpose a database was set up to hold all information required for the data reduction and analysis (see Figure 5.3). The database consisted essentially of four tables:

- *Observations* – Information extracted directly from the raw data headers
- *Calibrators* – Spectra and other information of about 800 N band calibrators, kindly provided by Roy van Boekel
- *Corr_fluxes* – Storage of all reduced and calibrated fringe track data
- *Photometries* – Storage of all reduced and calibrated single-dish data

The data reduction for a Large Programme source would then follow the outline given in Figure 5.3.

The best calibrator for any target observation (fringe track or photometry) was found automatically by using a weighting function that takes into account both the angular and temporal distance of the calibrator star from the science target. References for reasonable values for the relative weighting of these two distances are hard to find in the literature. At the Keck Interferometer it is assumed that either $t_0 = 1$ hour temporal distance or $\phi_0 = 15$ degree sky distance lead to an increase of variance in the calibration by a factor of two.[7] These values were used to determine the 'penalty' function p for a calibrator observation given a target source's position and time. p was defined as a function of time difference Δt and angular difference on the sky $\Delta \phi = \sqrt{(\Delta RA \cdot \cos \delta)^2 + (\Delta \delta)^2}$

$$p = \frac{\Delta t}{t_0} + \frac{\Delta \phi}{\phi_0} \tag{5.1}$$

---
[7] http://nexsci.caltech.edu/software/V2calib/wbCalib/index.html

The calibrator with the lowest value of p was chosen to calibrate the science observation. Normally these were simply the calibrators observed specifically for the science source. However, this simple function replaced the tedious procedure of manually matching target and calibrator observations for several hundred observations. Besides, for repeated calibrator – target – calibrator observations, it made sure that each target observation was matched with the closest calibrator observation.

5. The MIDI Large Programme: A statistical sample of resolved AGN tori

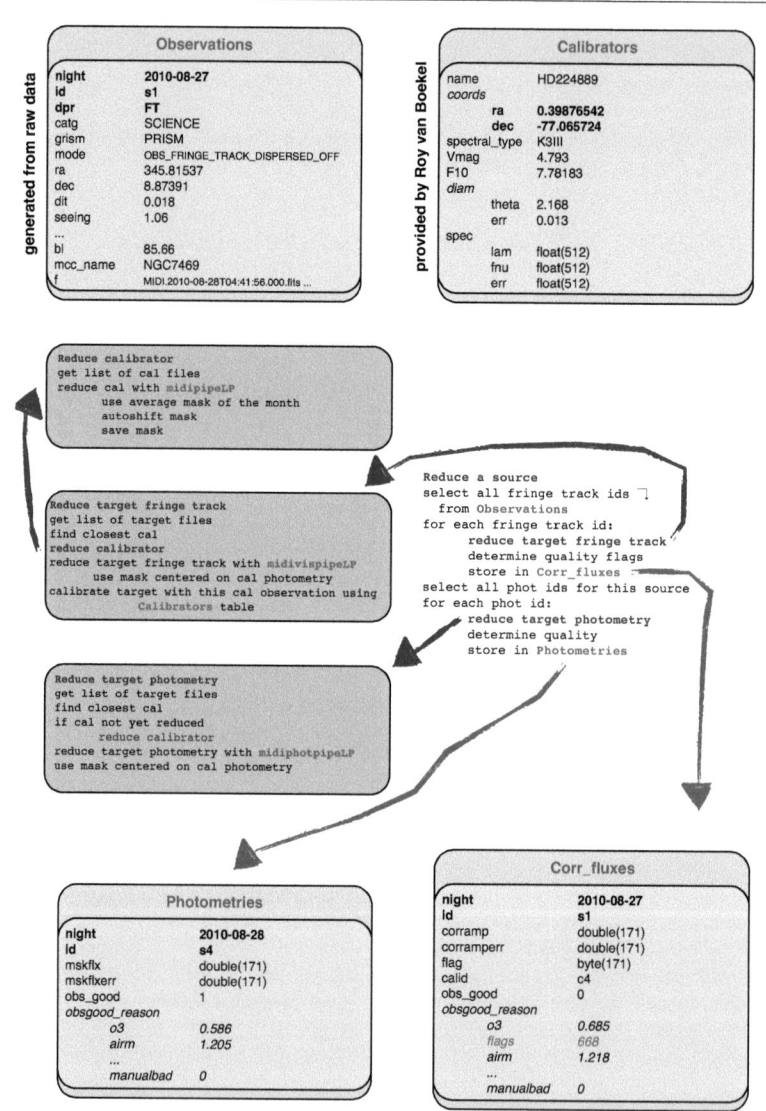

Figure 5.3.: Scripts (blue, in pseudocode) and selected fields of database tables (magenta) used for the LP data reduction, the scripts midipipeLP, midivispipeLP and midiphotpipeLP are a part of the data reduction with EWS that was described in Section 2.6. A combination of fields (printed in bold) served as unique identifiers of a dataset in a table.

5.2. Observations and Data Reduction

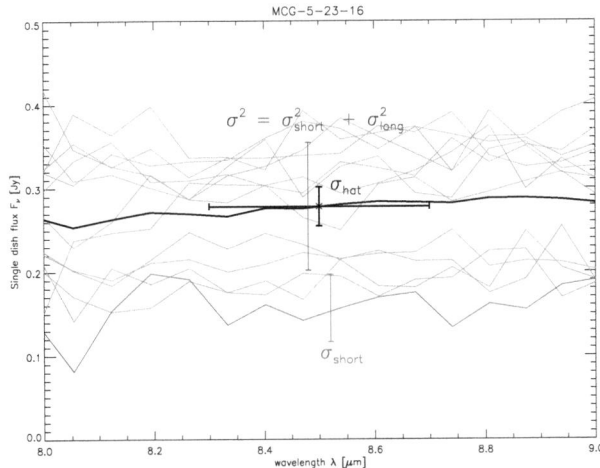

Figure 5.4.: Explanation of the error sources using the example of a set of MCG-5-23-16 single-dish observations (grey lines): A short-timescale error σ_{short} (red) can be determined from an individual observation, the total error measured on a set of observations (blue) is assumed to be the quadratic sum of this error and a long-timescale error σ_{long}. Using this decomposition, the error of the flux averaged over both all observations and a certain wavelength range (black, $\sigma_{hat} = \hat{\sigma}(\lambda \pm \Delta\lambda)$) can be determined. The blue and red error bars were slightly offset for readability. See text for details.

5.2.5. Uncertainties in the calibrated data

Single-dish spectra MIDI single-dish spectra of all but the strongest sources ($F_\nu \gg 10$ Jy) have large uncertainties. This affects all targets of the Large Programme. However, since we have taken $N_{\text{obs}} \gg 1$ such spectra per source in the course of the Programme, we are able to reduce the error (for most sources) significantly by using averaged fluxes. Yet when averaging both between observations and within one observation (i.e. over a wavelength range $\pm\Delta\lambda$), one must take into account that the uncertainty of the single-dish spectra has two main components that occur on different timescales and have to be treated differently in the averaging process (see Figure 5.4):

1. $\sigma_{\text{short}}(\lambda)$, an error that occurs on relatively short timescales ($\ll t_{\text{obs}} \approx 2$ min) and is probably dominated by photon noise. It can be estimated reliably by computing

5. *The MIDI Large Programme: A statistical sample of resolved AGN tori*

the variance[8] between subsets of one observation (see Section 2.7.3).[9]

2. $\sigma_{\text{long}}(\lambda)$, an error that is negligible for one observation but affects offsets *between* repeated observations. It is probably dominated by the imperfect background subtraction of single-dish observations (see Section 2.6.6) and it is reasonable to assume that this error follows a Gaussian distribution with zero mean.

Since we have taken multiple observations, we can determine both $\sigma_{\text{short}}(\lambda)$ (out of each individual observation) and the total error $\sigma(\lambda)$ (from the variance of multiple observations) empirically. Thus we can determine $\sigma_{\text{long}}(\lambda)$ under the assumption

$$\sigma^2(\lambda) = \sigma_{\text{short}}^2(\lambda) + \sigma_{\text{long}}^2(\lambda). \tag{5.2}$$

Figure 5.5 shows this decomposition for the five sources for which we have seven or more good single-dish spectra. It is evident that $\sigma_{\text{long}}^2(\lambda)$ (the blue curve) dominates in all sources except in MCG-5-23-16.[10]

Since eight of our thirteen sources do not have a sufficiently large number (≥ 7) of single-dish observations to derive $\sigma_{\text{long}}(\lambda)$ from the spread of the data, we use the average value of $\sigma_{\text{long}}(\lambda)$ from the more frequently observed sources (Figure 5.5) for the error determination of all sources. This basically sets a lower limit to the derived values of $\hat{\sigma}^2(\lambda \pm \Delta\lambda)$.

The error $\hat{\sigma}(\lambda \pm \Delta\lambda)$ of the flux averaged at a certain wavelength and over many observations is then given by

$$\hat{\sigma}^2(\lambda \pm \Delta\lambda) = \frac{1}{N_{\text{obs}}} \left(\frac{\langle \sigma_{\text{short}}^2 \rangle_{\text{obs},\lambda}}{N_\lambda} + \langle \sigma_{\text{long}}^2 \rangle_\lambda \right) \tag{5.3}$$

where $\langle \cdot \rangle_{(\text{obs},)\lambda}$ denotes a variance, averaged over (all observations and) $\lambda \pm \Delta\lambda$. N_λ is the number of bins in the wavelength range $\lambda \pm \Delta\lambda$.

If σ_{long} is really determined by imperfect background subtraction, it should be independent of source brightness (so that the relative error decreases). And if σ_{short} is dominated by shot noise, the SNR is $\propto \sqrt{N}$ where N is the number of counts on the detector. Between 9 and 11.5 μm the brightest source, IC 4329A ($F_\nu^{12.5\mu m} \approx 1$ Jy), indeed has a larger total error than the other sources – but the lowest $\sigma_{\text{short}}(\lambda)$. We conclude that – in the range of fluxes studied here – other factors than the source flux seem to determine the errors and thus justify the approximation of applying the same value of σ_{long} for all sources.[11]

---

[8]The variance is defined as $\sigma^2(\lambda) = \langle \sigma^2(\lambda) \rangle - \langle \sigma(\lambda) \rangle^2$, where $\langle \cdot \rangle$ denotes the arithmetic average.

[9]Photon noise follows Poisson's statistics, but in this case it can be well approximated by Gaussian statistics due to the large number of counts (even for weak sources).

[10]This is the only type 2 source in this sample and the only source for which the adaptive optics was set to off-target guiding. A correlation between AO guiding mode and source quality was not further investigated.

[11]The flux errors for much brighter sources, such as the Circinus galaxy or NGC 1068 (see Section 5.4.1.1) show that single-dish errors in MIDI remain a nuisance even for moderately bright sources ($F_\nu \approx 10$ Jy) where the relative errors are still comparable to the weak LP targets.

5.2. Observations and Data Reduction

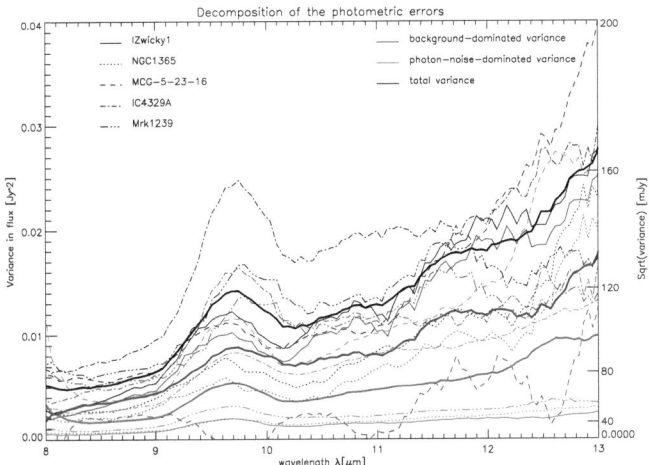

Figure 5.5.: Decomposition of the error in MIDI single-dish spectra of weak sources: Total observed variance $\sigma^2(\lambda)$ (black), short-time fluctuations $\sigma^2_{\text{short}}(\lambda)$ (red), long-time fluctuations $\sigma^2_{\text{long}}(\lambda)$ (blue), see text for details. The sum of the red and the blue curve are the black one. Thin lines with various linestyles: individual sources (indicated in the plot); thick lines: average values for all sources. In the average curves, a local maximum can be seen in the region of telluric Ozone absorption. A boxcar-smoothing with a width of about 0.2 μm has been applied to all curves for readability.

Correlated flux spectra Correlated fluxes are less affected by an uneven background since most of the background is removed through high-pass filtering (see Section 2.6.2). Experiments with repeated observations of identical (u, v) points (Section 2.7.4) suggest that, for correlated fluxes, $\sigma_{\text{long}} \lesssim \sigma_{\text{short}}$. The errors in correlated fluxes are therefore simple errors of the mean, i.e. the flux averaged over the given wavelength regime.

Systematic errors Systematic errors have been studied specifically for the LP dataset. The tests and results are described in Section 2.7.4.

5.3. Results

5.3.1. (u, v) coverages

The resulting (u, v) coverages are displayed in Figure 5.6.

For the "detailed map sources" NGC 1365, MCG-5-23-16 and IC 4329 A, continuous fringe tracks[12] on at least one baseline were planned but only partially achieved for NGC 1365 and MCG-5-23-16. In IC 4329 A no continuous fringe track was observed but a good overall coverage of the (u, v) plane was achieved.

For the sources of the "extended snapshot" sample, six (u, v) points were planned so that two or three position angles would be sampled by two or three baseline lengths each. It was attempted to span a right angle between two of these directions in order to detect elongation.

For two sources, the desired coverage has been achieved (I Zwicky 1 and LEDA 17155). For the others, gaps remain, mostly as a result of weather loss. Follow-up observations are scheduled to compensate for this loss (see Section 5.7).

5.3.2. Correlated flux and single-dish spectra

All correlated flux and single-dish spectra can be found in the original publication of this Ph.D. thesis (Burtscher 2011).

The single-dish spectra are colored by year to show that the variability in these fluxes is less than the errors of the individual photometries. In other words: To the accuracy of the single-dish uncertainties ($\approx 30\%$), we can exclude flux variability in the mid-infrared between the respective observations.

Neither the single-dish nor the correlated flux spectra show any spectral lines (not even the often seen [Ne II] 12.81 μm forbidden line, indicative for star forming regions). The only feature present in some of the spectra is that of silicates. It can be seen most clearly in emission in I Zwicky 1 and in Mrk 1239 and in absorption in LEDA 17155, NGC 3281 and in NGC 5506.

5.3.3. Visibilities on the (u, v) plane

In the original publication of this Ph.D. thesis (Burtscher 2011), the visibilities are displayed on a (u, v) plane (and again colored by year) to give an overview of the level and range of the visibilities in the targets. From these plots, it can be seen that there are no signs for elongation in any of the targets, with the possible exception of MCG-5-23-16.

---

[12]A continuous fringe track is a dense sampling of the target (u, v) plane, only intercepted for one calibrator observation per hour.

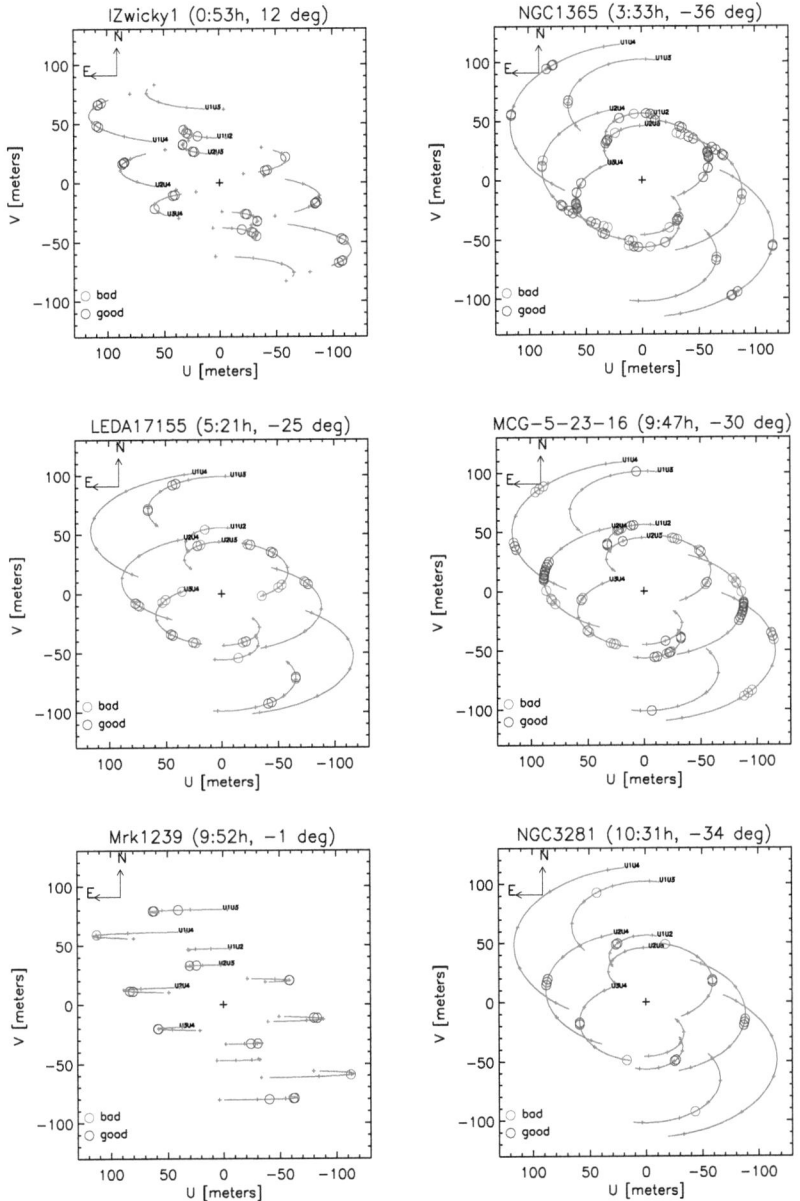

Figure 5.6.: (u,v) coverages of LP targets. Observations flagged as good according to the above mentioned criteria are displayed in blue, others in red. The diameter of the circles corresponds to 8 m, the diameter of the VLT UTs. Small grey crosses denote Hour Angles = -4, -2, 0, 2, 4. (u,v) plane tracks are followed CCW if $\delta < 0$, CW if $\delta > 0$. *Continued on next page*

5. The MIDI Large Programme: A statistical sample of resolved AGN tori

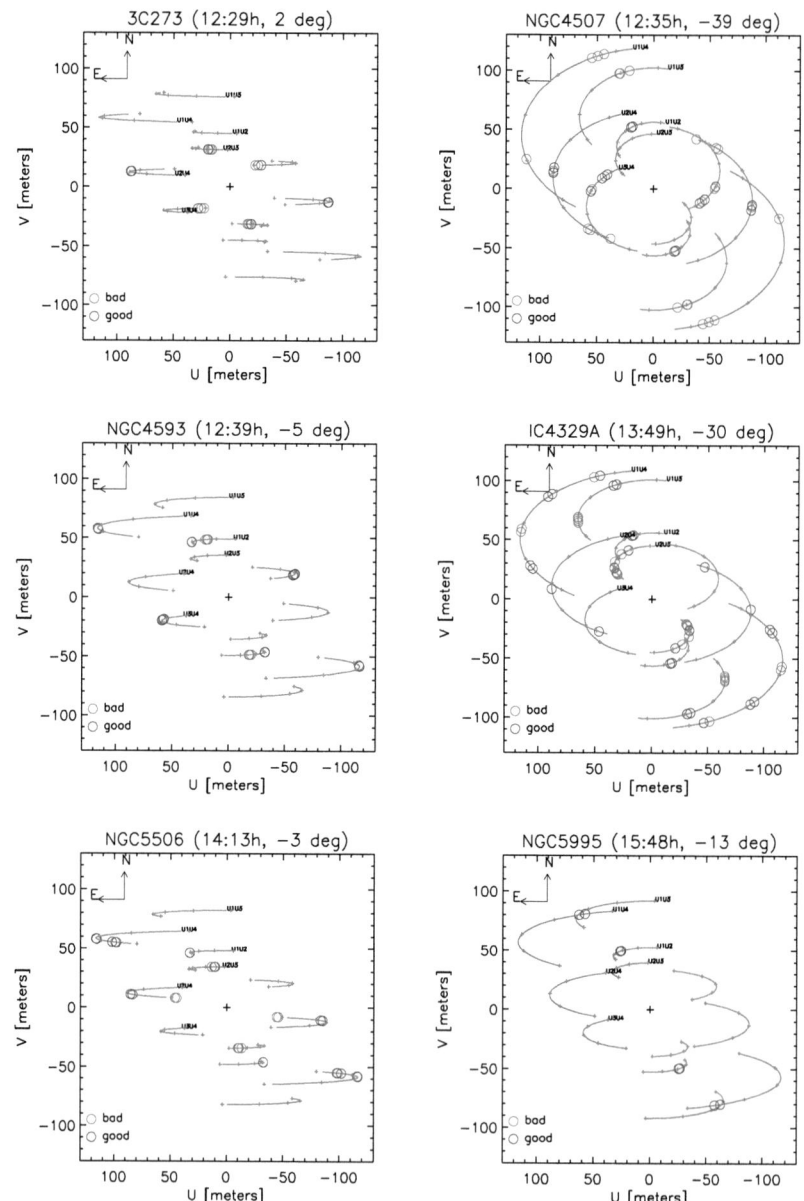

Figure 5.6.: — *Continued*

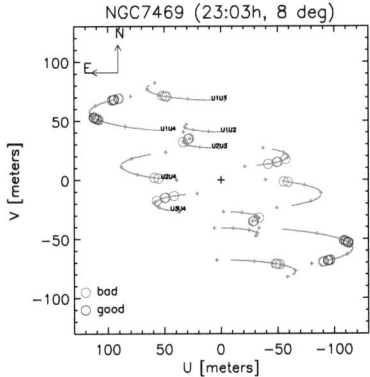

Figure 5.6.: — *Continued*

5.4. Radial visibility models

With phase-less visibilities (correlated fluxes) on sparsely sampled (u, v) planes, *images* cannot be reconstructed. We therefore compare the observed fluxes to models (see Section 2.2.6). The goal of these first model fits to the LP targets is to derive reasonable estimates (with errors) of the size and flux of the resolved emitter, if it exists, and an estimate of the unresolved flux. To this end we are looking for a decrease of correlated flux with increasing baseline length and fit a Gaussian profile to it.

We fit fluxes and not visibilities (see also discussion in Section 2.7.1). This way the (relatively large) error of the single-dish flux (taken as the 'zero baseline' point) is cleanly separated from the errors of the correlated fluxes.

Since no pronounced dependence of visibility on position angle is visible from the visibility plots on the (u, v) plane (Burtscher 2011), circular symmetry is assumed for all model geometries and we can reduce our co-ordinate system to one dimension, the spatial frequency in radial direction BL_λ.

We employ the following radial models for the (correlated) fluxes and fit the parameters of the models at three wavelengths: 8.5 ± 0.2 μm, 10.5 ± 0.2 μm and 12.5 ± 0.2 μm, avoiding regions of low SNR and the telluric ozone feature, but probing the region of silicate absorption with the 10.5 μm fit.

a) **Point source + Gaussian** This model consists of an unresolved point source with flux contribution F_p (in mJy) and a (partly) resolved Gaussian with flux F_g (in mJy) and FWHM Θ. It seems to be an adequate fit for many of our observations.

$$F_\nu(BL_\lambda, F_p, F_g, \Theta) = F_p + F_g \cdot \exp\left(-\frac{(\pi \Theta BL_\lambda)^2}{4\ln 2}\right) \quad (5.4)$$

This model has been applied e.g. for I Zwicky 1 (Figure 5.7).

b) Point source + lower limit for over-resolved Gaussian In sources where the correlated flux may be lower than the single-dish flux, but no decrease of correlated flux with increasing baseline length is visible, the source is assumed to consist of a point source and an over-resolved Gaussian component to explain the difference between the correlated and single-dish fluxes. In this case, only a lower limit to the FWHM of this Gaussian component Θ (with flux F_g) is given by requiring that the over-resolved component has dropped in flux at least to the level of the standard deviation of the correlated fluxes σ_c at the lowest observed spatial frequency $BL_{\lambda,\min}$, i.e.

$$F_\nu(BL_{\lambda,\min}, F_p, F_g, \Theta) \lesssim F_p + \sigma_c \tag{5.5}$$

Substituting Equation 5.4 and rearranging gives a limit for Θ:

$$\Theta \gtrsim \frac{2\sqrt{2 \ln F_g/\sigma_c}}{\pi BL_{\lambda,\min}} \tag{5.6}$$

This model has been applied e.g. for NGC 3281 and is further explained in Figure 5.13.

c) Point source + large Gaussian + small Gaussian In partly resolved sources where a shallow decrease of correlated flux with increasing BL_λ can be seen, we will consider to add a second, small, Gaussian component (flux F_{g2}, FWHM Θ_2) to the fit. The flux and Gaussian FWHM of the well resolved emitter are labeled F_{g1} and Θ_{g1} in this model.

$$F_\nu(BL_\lambda, F_p, F_{g1}, \Theta_{g1}, F_{g2}, \Theta_{g2}) = F_p + F_{g1} \cdot \exp\left(-\frac{(\pi\Theta_1 BL_\lambda)^2}{4\ln 2}\right) + F_{g2} \cdot \exp\left(-\frac{(\pi\Theta_2 BL_\lambda)^2}{4\ln 2}\right) \tag{5.7}$$

This model has been applied for IC 4329 A (Figure 5.18) and NGC 5506 (Figure 5.19).

5.4.1. Results

I Zwicky 1 – a very luminous narrow-line Seyfert 1 galaxy This type of AGN is thought to deviate from the $M_{BH} - \sigma$ relation in the sense that they have 'undermassive' black holes and therefore no strong broad emission lines (Komossa 2008). However, I Zw 1 is also sometimes called a quasar (PG 0050+124, Weedman et al. 2005).

It has a nuclear starburst on kpc scales and a possibly connected circum-nuclear molecular ring (Schinnerer et al. 1998). It is believed that I Zwicky 1 is undergoing a merger with the nearby companion galaxy that is marginally detectable in Figure 5.1 to the west (Scharwächter et al. 2003).

I Zw 1 is well studied in many wavelength regimes: In the X-Rays, different modes of short-term variability are observed from which a two component accretion-disk corona is implied (Gallo et al. 2007). It is also a popular target for observations in the UV for its

5.4. Radial visibility models

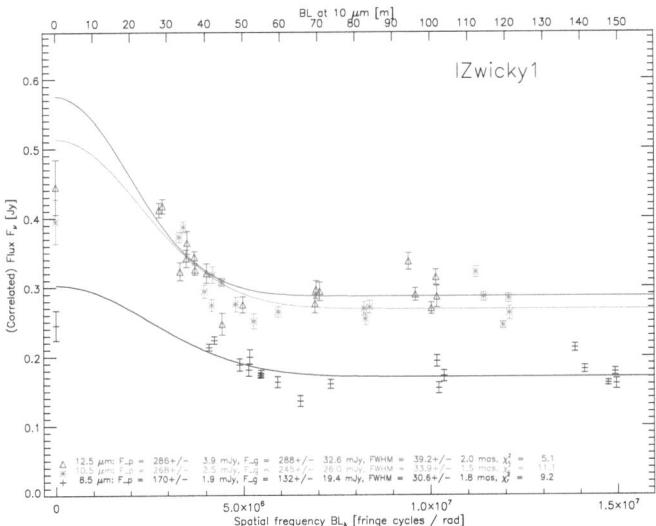

Figure 5.7.: Radial (1D) visibility model for I Zwicky 1. The single-dish ("0-baseline") and correlated fluxes are plotted as a function of spatial frequency for three wavelengths, together with the best fit for a Gaussian + point source model, see text for details. The best fitting parameters are given in the plot.

rich Fe II emission spectrum[13]. In the infrared, I Zw 1 is found to have "strong silicate emission and no PAH or emission lines" (Weedman et al. 2005). A dedicated review about this source does not seem to exist, but the introduction of Scharwächter et al. (2007) gives a concise overview of this source.

In our radial models, I Zwicky 1 is the perfect case for the point source + Gaussian model (a). In all three wavelengths, a clear decrease in correlated flux is seen with increasing spatial frequency. At spatial frequencies $\gtrsim 6 \cdot 10^6$, the flux is at the point source level and does not decrease any more with increasing BL_λ. The fits predict larger single-dish fluxes than observed. This is a consequence of the small errors of the correlated fluxes in comparison with the single-dish fluxes. In this and in the following fits, we decided not to require the fits to meet the single-dish fluxes but to respect the derived statistical errors.

The best fitting values for I Zwicky 1 and for all other sources are given in the respective Figure and in Table 5.2 on page 147.

---

[13] It is hoped to develop an understanding of the Fe II emission mechanism to use it as a diagnostic for BLRs (Bruhweiler & Verner 2008).

Figure 5.8.: The "Great Barred Spiral Galaxy" **NGC 1365** as seen in the infrared ($YJHK$ composite) by HAWK-I at the VLT. Image credit: ESO/P. Grosbøl

5.4. Radial visibility models

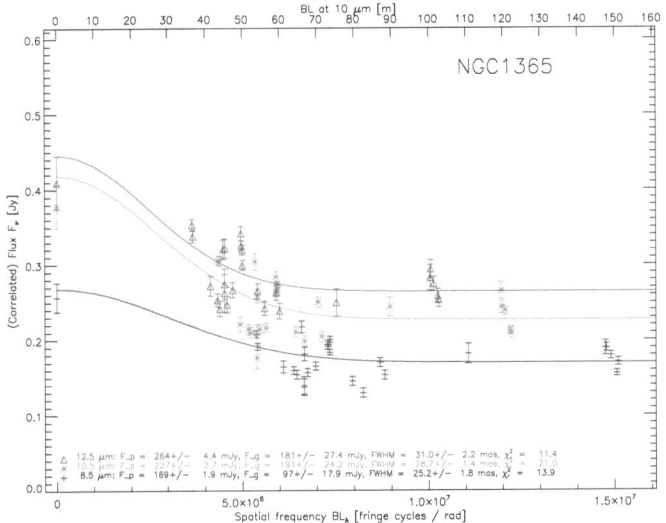

Figure 5.9.: Same as Fig. 5.7 but for **NGC 1365**

NGC 1365 – the one with black hole eclipses The supergiant barred galaxy NGC 1365 is the closest (and maybe the most beautiful, Figure 5.8) of the 13 targets observed in the Large Programme. It exhibits a wide variety of nuclear activity and has been studied, partly because of its prominent bar, in great detail. A radio jet emanates from the optical nucleus and is visible up to ~ 500 pc from the center (Sandqvist et al. 1995) and CO observations have revealed a giant molecular torus of over 1 kpc in diameter (Sandqvist 1999). An excellent review has been given by Lindblad (1999).

Fits to the optical spectrum of NGC 1365 show that a broad component of Hα emission is required and that it is probably a Seyfert 1 galaxy seen through $A_V = 3$ absorption (Veron et al. 1980). However, the optical classification is somewhat vague as some classify it as a Seyfert 1.5 (i.e. 'type 1', Lindblad 1999) and others as a Seyfert 1.8 (i.e. 'type 2', Maiolino & Rieke 1995; Véron-Cetty & Véron 2006). So much seems to be clear: This source is between the type 1 and the type 2 class – and it may even change class due to intrinsic variations. This has been demonstrated very impressively from X-Ray observations in which Risaliti et al. (2007) have seen a dramatic change in its X-Ray behaviour: The source changed from Compton-thin to reflection-dominated back to Compton-thin in just four days. This was interpreted as a Compton-thick cloud close to the broad line region that, on its orbit around the nucleus, eclipsed the broad line region while in our line of sight. The deduced size of these X-Ray absorbing clouds, $r \approx 10^{12}$ m $\approx 0.1\,\mu$ arcsec,

5. The MIDI Large Programme: A statistical sample of resolved AGN tori

is clearly out of reach for MIDI, however.

In the mid-IR, [S IV] and [Ne II] were detected by ISO observations, but not in observations with a smaller aperture (Siebenmorgen et al. 2004). These TIMMI2 observations also showed no sign of silicate absorption.

In our radial models, NCG 1365 is, besides I Zwicky 1, the other case where both the decrease of correlated flux and the point source flux level can be clearly seen, i.e. model (a) is well defined. However, the scatter is much larger leading to much larger χ_r^2 values than for the distant I Zwicky 1. A possible interpretation for some of the scatter is discussed in Section 5.5.2.

In the GTO study (Tristram et al. 2009) this source was found to be partially resolved (confirmed by the new data) and possibly elongated. This cannot be confirmed, see the visibilities on the (u,v) plane in Burtscher (2011).

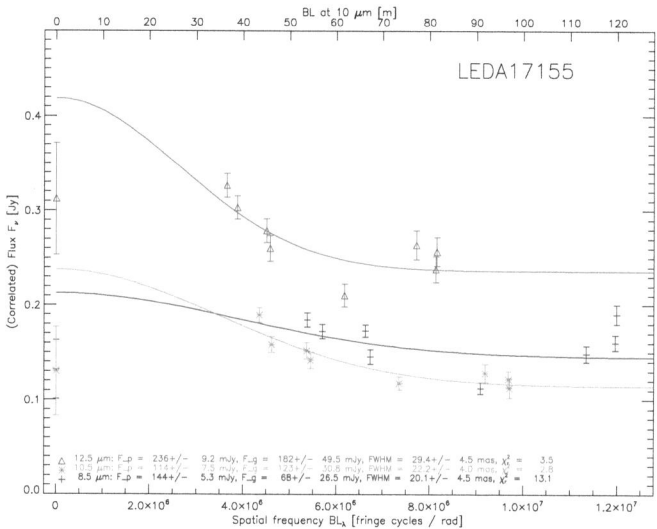

Figure 5.10.: Same as Fig. 5.7 but for **LEDA 17155 / IRAS 05189-2524**

LEDA 17155 / IRAS 05189-2524 – a ULIRG With $\log L(8 - 1000 \mu m)/L_\odot = 12.10$, LEDA 17155 (= IRAS 05189-2524) is an ultraluminous infrared galaxy (ULIRG, Sanders et al. 1988). From optical spectroscopy, it was classified as a Seyfert 2 galaxy by Veilleux et al. (1995) and as a 'hidden Seyfert 1' galaxy by Young et al. (1996) due to the detection of a broad Hα line in polarized light. Deep HST/WFPC observations ($\approx B$ and I bands)

5.4. Radial visibility models

reveal clumps (probably star-forming regions) as well as "loops" and "horns" that are probably tidal features (Surace et al. 1998). In a recent X-Ray observing campaign, Suzaku observations did not show any short-term variability (like in NGC 1365), but revealed a change of flux compared to previous observations with other X-Ray satellites (Teng et al. 2009).

In the mid-IR, Spoon et al. (2002) identified a spectral feature due to water ice at 6 μm and Siebenmorgen et al. (2004) estimated $A_V \approx 12$ mag from the silicate absorption feature.

The radial models at 10.5 μm and 12.5 μm provide good fits to the MIDI data. At 8.5 μm the source is essentially unresolved.

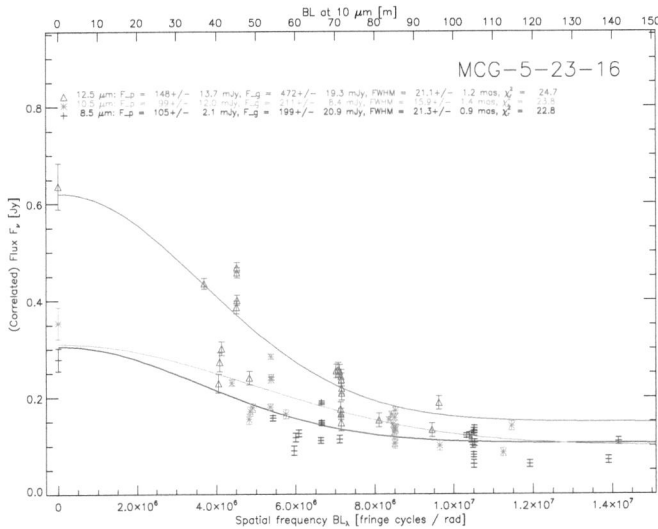

Figure 5.11.: Same as Fig. 5.7 but for **MCG-05-23-16**

MCG-05-23-16 This galaxy probably has a hidden type 1 nucleus as suggested both by its original classification by Veron et al. (1980) who found a broad component for the fit to the Hα line and by the current classification as a type 1i (Véron-Cetty & Véron 2006) meaning that a broad Paschen β line is detected. In the X-Rays, MCG-05-23-16 shows an unusual Fe Kα line profile that is difficult to fit with standard accretion disk models (Weaver et al. 1997).

In this relatively nearby source, again a clear decrease in fluxes with BL_λ is seen (model a). At two spatial frequencies, $4 \cdot 10^6$ and $7 \cdot 10^6$ for the 12.5 μm fit, the measured fluxes show a large scatter around the best-fit value. While the spatial frequency is nearly the same for all of these values, the position angle changes by about 10°. A possible dependence of correlated flux on PA is discussed in Section 5.5.2 and Figure 5.26.

Tristram et al. (2009) found the source partially resolved.

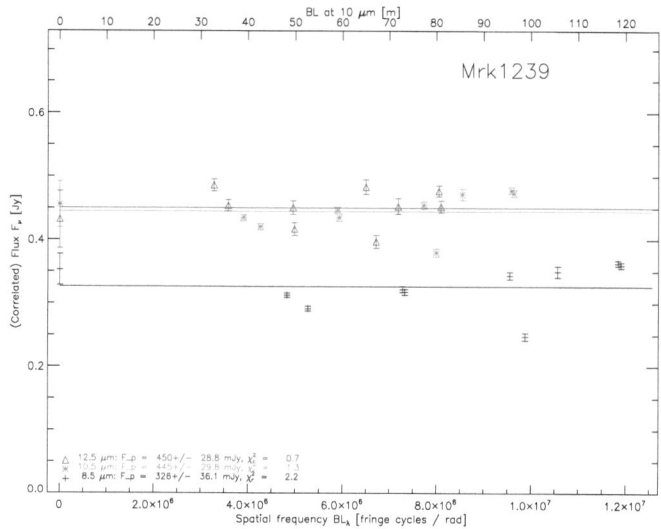

Figure 5.12.: Same as Fig. 5.7 but for **Mrk 1239**

Mrk 1239 In an infrared spectrum taken with the Infrared Telescope Facility (IRTF), Rodríguez-Ardila & Mazzalay (2006) find an unusually strong bump of emission peaking at 2.2 μm that is well fitted by a 1200 K blackbody. In the mid-infrared, no PAH emission is seen (indicating the absence of star formation) – probably because the infrared flux of the whole galaxy is dominated by emission from the AGN (Reunanen et al. 2010).

In this source, all MIDI correlated fluxes are on the same level as the single-dish fluxes: Mrk 1239 is unresolved in the mid-IR on $\approx$ 10 mas scales, confirming the result found from the GTO observations (Tristram et al. 2009). In this case, a Gaussian component is not required and we defined the point source flux and error as the average and rms error of the average, respectively.

NGC 3281 This Seyfert 2 galaxy is notable for its highly obscured nucleus and its relatively small ratio of visual extinction A_V to absorbing hydrogen column density N_H

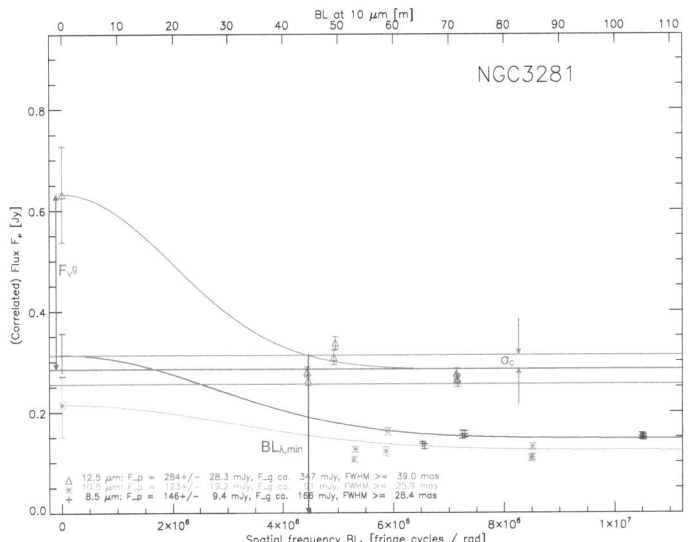

Figure 5.13.: Same as Fig. 5.7 but for **NGC 3281**. Model b is explained using the example of the 12.5 μm fit. A lower limit for the size of the over-resolved component with flux F_g is derived by requiring the model visibility curve (red) to have dropped to the point source level + the spread in point source flux σ_c at the lowest spatial frequency $BL_{\lambda,\min}$. In other words: the red curve is required to intersect the upper yellow curve at $BL_{\lambda,\min}$.

(1/50 the Galactic value) as found by Simpson (1998) in a near-IR and X-Ray study of this galaxy. This finding was explained by Vignali & Comastri (2002) with a special geometry of the obscuring material, although this must not necessarily be the case since a large range in A_V/N_H, and especially values lower than the Galactic value, have been found in AGNs by Maiolino et al. (2001). With respect to anomalous A_V/N_H values, see also the discussion for the 'unabsorbed Seyfert 2' NGC 5995. While there is no direct evidence of a broad line region in this galaxy (it is classified as a Seyfert 2.0), Storchi-Bergmann et al. (1992) argue that it most probably has a hidden type 1 nucleus as derived from a well-defined biconical ionization region and emission line models.

In NGC 3281 the point source level is adequately well defined by the MIDI correlated fluxes. By requiring that the contribution from the Gaussian component is lower than the rms of the correlated fluxes (model b), we can constrain a lower limit to the size of the Gaussian component.

5. The MIDI Large Programme: A statistical sample of resolved AGN tori

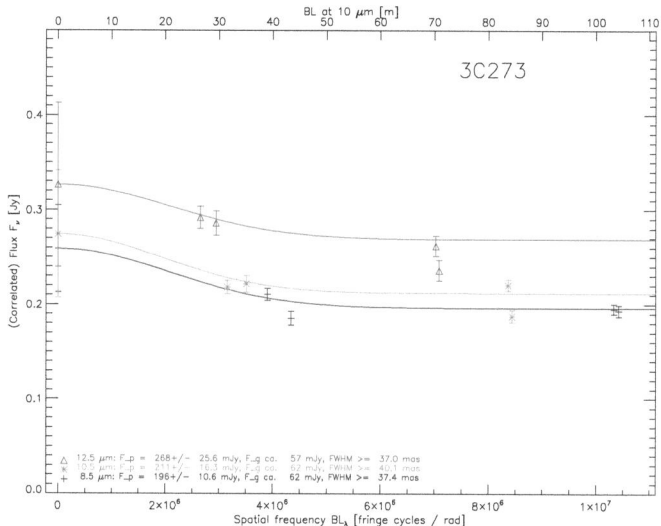

Figure 5.14.: Same as Fig. 5.7 but for **3C 273**

3C 273 – an extremely luminous, distant source The most distant source in the LP is the quasar 3C 273. It is also the most luminous source accessible with MIDI. With a luminosity $\nu L_\nu \approx 4 \cdot 10^{38}$ W (at $\lambda = 12$ μm), it is about an order of magnitude brighter than the second most luminous galaxy in our sample. At the cosmological redshift of $z = 0.158$ ($D_a \approx 546$ Mpc) it is the most distant object that can be studied with MIDI.

Strictly, it was not part of the Large Programme, because at the time of proposal submission, it was scheduled for observations on a DDT programme (282.B-5071). Unfortunately, that programme was only partly executed due to bad weather and timing constraints. During the LP time, 3C 273 was successfully observed in 2010-01-30 as a backup target.

3C 273 is one of the most well-studied quasars and was the first to be identified with a cosmological source due to its high redshift by Schmidt (1963). The literature for 3C 273 is rich, an overview is given in the review by Courvoisier (1998).

In the infrared, a bump at 3 μm had already been noticed by Neugebauer et al. (1979) and was attributed to the presence of hot dust. On the other hand, this radio-loud quasar has a prominent jet that shines brightly from radio to gamma rays with contributions also in the infrared and shows variability across the electromagnetic spectrum (Soldi et al. 2008). The infrared quiescent flux level has been interpreted as being due to dust emission, however (Robson et al. 1993). This interpretation was confirmed by an observation in a

5.4. Radial visibility models

historic minimum in the sub-millimeter emission of 3C 273's jet where Türler et al. (2006) identified three thermal components of dust at various temperatures ($T \approx$ 40, 250, 1300 K). Apart from that, Hao et al. (2005) presented Spitzer spectra that showed silicate emission features at 10 and 18 μm in 3C 273, requiring the presence of hot dust.

From early MIDI data, the source appeared to be possibly resolved. Tristram et al. (2009) gave two *upper* limits for the size (<67 pc / < 108 pc). The reasoning for this limit was that the correlated flux appeared significantly different at the two baselines observed. This flux difference could not be reproduced with the more extensive LP dataset. Instead, evidence for a decrease of correlated flux with increasing baseline length is marginal and the new observations are compatible with an over-resolved Gaussian that, at the distance of 3C 273, would have to be $\gtrsim$ 100 pc. Most of the emission is unresolved, i.e. on scales $\lesssim \lambda/3BL \approx 7$ mas ≈ 20 pc.

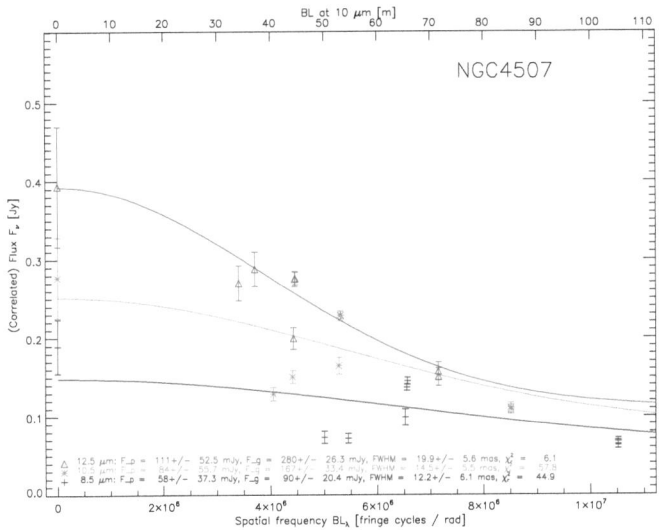

Figure 5.15.: Same as Fig. 5.7 but for **NGC 4507**

NGC 4507 This is another hidden Seyfert 1 galaxy as seen by broad lines in polarized flux (Moran et al. 2000). This Compton-thin, X-Ray bright galaxy has been studied in detail with XMM-Newton and Chandra by Matt et al. (2004). In the mid-infrared the source is unresolved with single-dish observations (Horst et al. 2009).

In NGC 4507, only at 12.5 μm is a clear decrease of correlated flux seen with increasing spatial frequency. But is it real? The two lowest spatial frequency fluxes at 12.5 μm are,

5. The MIDI Large Programme: A statistical sample of resolved AGN tori

also relatively, higher than at other wavelengths. Their errors, given in the fit in Figure 5.15, are the largest correlated flux errors in this source. These fluxes are probably overestimated due to the spurious "emission feature" at $\lambda \approx 12.5 \mu m$ seen in the spectra (Burtscher 2011). Reassuringly, however, the fitted FWHM sizes would not change dramatically if these two points were placed at lower fluxes. Also, all sizes are compatible with each others within the errors.

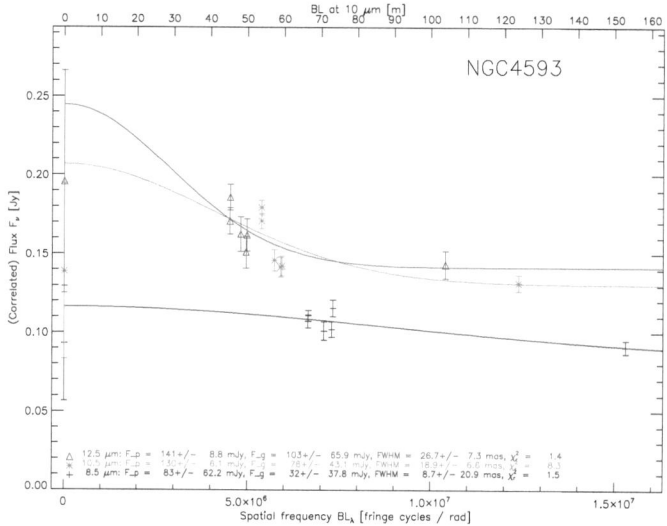

Figure 5.16.: Same as Fig. 5.7 but for **NGC 4593**

NGC 4593 The Seyfert 1 galaxy NGC 4593 is an example of a so called "pseudobulge galaxy" (Kormendy et al. 2006) whose dense nuclear region might be mis-interpreted as a merger-built bulge but is actually thought to be the product of secular evolution. NGC 4593 is a popular target in reverberation-mapping campaigns due to its emission line variability (e.g. Greene et al. 2010). In Spitzer IRS spectra, the 9.7 μm silicate emission feature is seen, peaking at ≈ 10.5 μm (Horst et al. 2009).

NGC 4593 is among the weakest of all the LP targets with correlated fluxes between 100 and 200 mJy. Nevertheless, a clear decrease of correlated flux with baseline length is seen at all wavelengths, with the most prominent drops at 10.5 μm and 12.5 μm. At these wavelengths, the point source is well defined and the uncertainty in F_g is dominated by the large errors of the single-dish fluxes. At 8.5 μm the correlated flux drop is small

5.4. Radial visibility models

and the best fit is found for a very small Gaussian (with shallow decrease of correlated flux). The point source flux is not well defined at 8.5 μm.

Figure 5.17.: BVRI composite image taken at the VLT/FORS2 showing the interacting galaxy group IC 4329 A (left) + IC 4329. Is the activity in IC 4329 A triggered by the nearby elliptical galaxy? X-Ray images show a bridge of hot diffuse gas between the two galaxies (Read & Pietsch 1998). Image Credit: ESO/M. Mejias

5. The MIDI Large Programme: A statistical sample of resolved AGN tori

IC 4329A The X-Ray bright edge-on Seyfert 1 galaxy IC 4329 A (Figure 5.17) is found near the center of Abell cluster A3574. Its variability has been studied (with little success) in the optical (Peterson et al. 2004) and with more success in the X-Rays (Markowitz 2009). The latter derived a black hole mass of $1.3^{+1.0}_{-0.3} \times 10^8 M_\odot$ from an empirical relation linking it to the break in the power spectral density function and the bolometric luminosity of the galaxy. In the mid-IR, Siebenmorgen et al. (2004) found the [S IV] and the [Ne II] line in TIMMI2 spectra. From the GTO data, the source appeared unresolved (Tristram et al. 2009).

In the LP data, IC 4329A is one of the two sources (besides NGC 5506) where the plot of correlated fluxes versus baseline (Figure 5.18) gives the impression of two resolved components: Clearly, there must be a very large ($\Theta \gtrsim 20\ldots40$ mas, depending on wavelength) component to explain the low level of correlated fluxes with respect to the single-dish fluxes. But when looking closer, one can also see a more shallow decrease of flux towards longer baselines – an indication for a second, much smaller and only partially resolved, Gaussian component?

Not really: Adding two parameters (F_{g2}, Θ_2) to accommodate for this component (i.e. moving from model a to model c), does not improve the goodness of fit as determined by the value of χ^2_r. While the absolute value of χ^2 decreases, the lower number of degrees of freedom leads to actually larger values of χ^2_r than for the simpler two component model. The χ^2_r values for model a are 27.0, 53.6 and 46.3 for the fits to the 8.5 μm, 10.5 μm and 12.5 μm data, respectively. In model c, $\chi^2_r = 27.9$, 54.1 and 54.1, respectively. Since our main aim here is to give reliable estimates of the sizes, we actually employ the much simpler model b and give a robust lower limit for the size of the resolved emitter. A large scatter remains around the point source flux level.

The fit with the two components as well as the more robust estimate of the extended component's size are displayed in Figure 5.18.

5.4. Radial visibility models

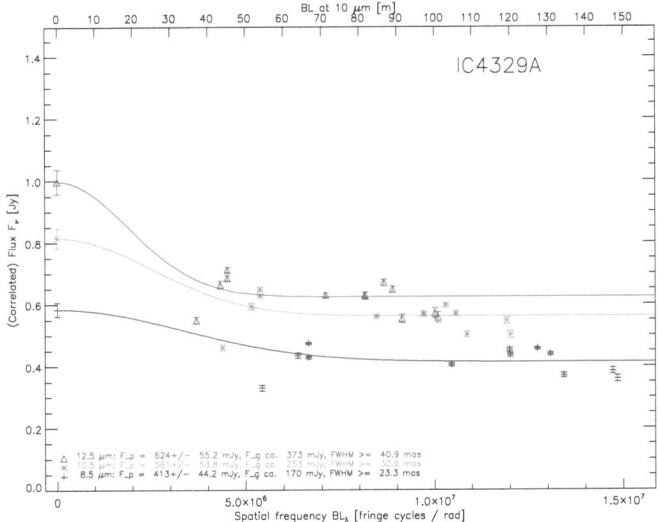

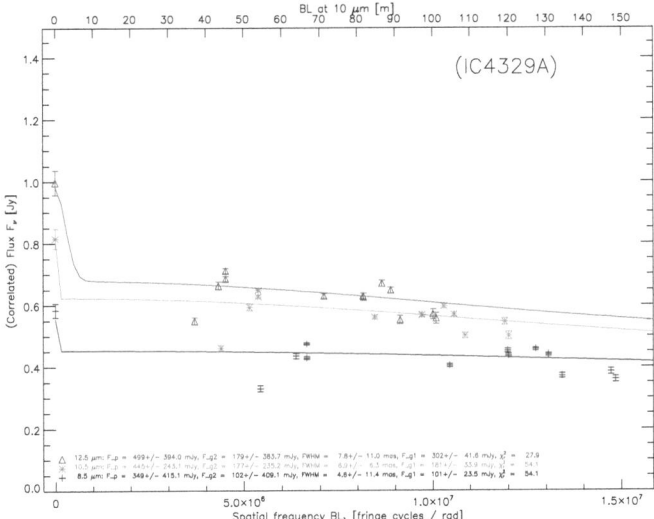

Figure 5.18.: Same as Fig. 5.7 but for **IC 4329 A**. Top panel: Model b. The size of the large Gaussian is not constrained in model c (lower panel) and only a limit can be given (model b). See text for details.

5. The MIDI Large Programme: A statistical sample of resolved AGN tori

NGC 5506 – an obscured Narrow Line Seyfert 1? Nagar et al. (2002) are convinced to have found in NGC 5506 the first case of an optically-obscured Narrow-Line Seyfert 1 (NL-Sy1): From IR spectroscopy they find permitted lines that, according to them, can only originate from the BLR. The FWHM of these lines, however, is < 2000 km/s admitting NGC 5506 indeed into the club of NL-Sy1 galaxies. NGC 5506 hosts a relatively luminous H_2O megamaser (Braatz et al. 1994; Henkel et al. 2005). Guainazzi et al. (2010) find a broad Fe Kα line in this galaxy and speculate that reprocessing of this line occurs in the AGN torus.

In this edge-on galaxy (see Figure 5.1), the nucleus is obscured by $A_V \approx 15$ mag as determined from the pronounced silicate absorption feature (Siebenmorgen et al. 2004; Burtscher 2011). From the unresolved nucleus at near-infrared wavelengths, Prieto et al. (2010) derive an upper limit to the FWHM size of the source of 0.1 arcsec (13 pc, FWHM). As in Mrk 1239, the IR flux of this galaxy originates almost exclusively from the nucleus (Reunanen et al. 2010).

NGC 5506 is the other candidate for a second, smaller, Gaussian component. In this case, a fit to model c leads to reasonably well defined values for the point source and the small (partially resolved) Gaussian component (see the two plots for this source in Figure 5.19). The over-resolved Gaussian component is not well defined. However, the χ_r^2 values for model a are not much different and we *choose* to use the parameter values of the simpler model a for the following discussion.

5.4. Radial visibility models

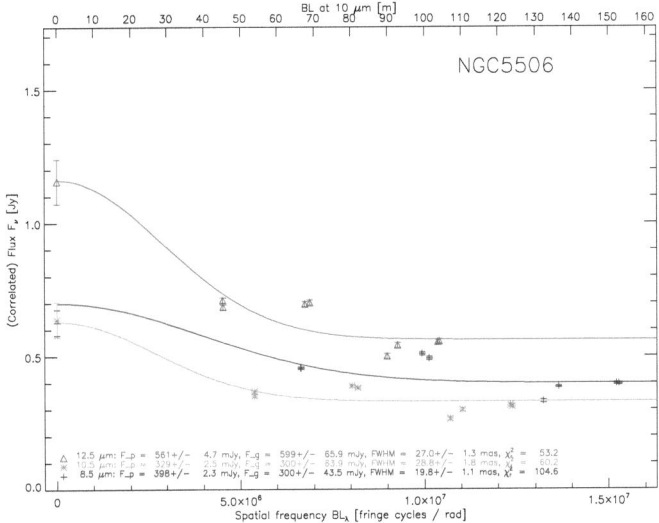

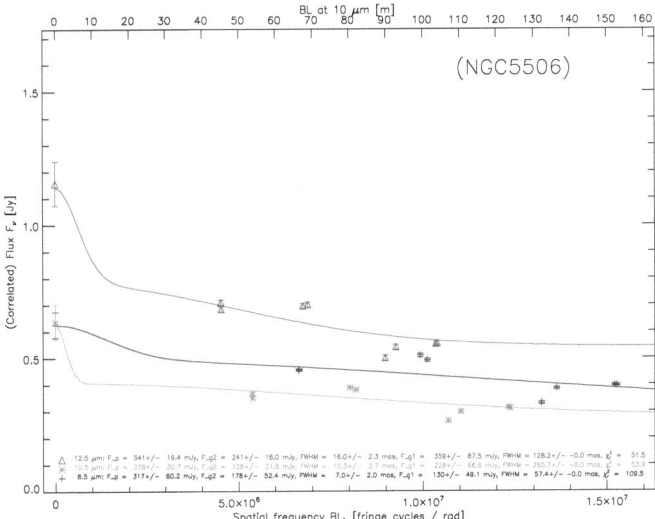

Figure 5.19.: Same as Fig. 5.7 but for **NGC 5506**. In model c (lower panel), the size of the large Gaussian component is not constrained, for a limit see model a (top panel).

5. The MIDI Large Programme: A statistical sample of resolved AGN tori

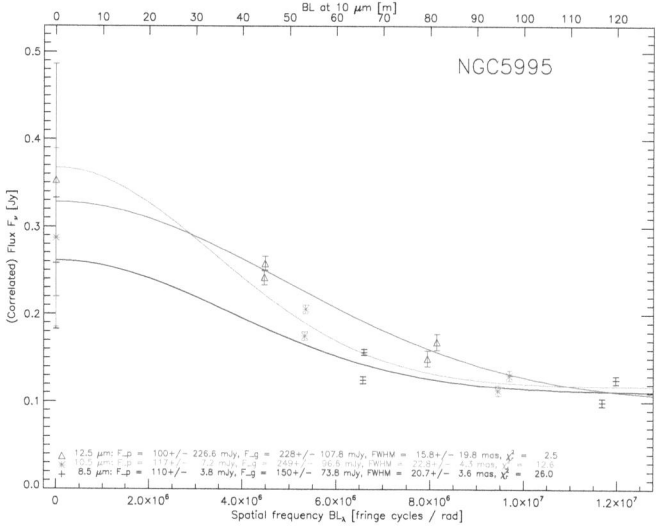

Figure 5.20.: Same as Fig. 5.7 but for **NGC 5995**

NGC 5995 – an unabsorbed Seyfert 2 NGC 5995 is among the few (4%, Risaliti et al. 1999) Seyfert 2 galaxies that have an X-Ray absorbing column $< 10^{22}$ cm^{-2} (a so called "unabsorbed Seyfert 2 galaxy"). Panessa & Bassani (2002) argue that the observed X-Ray absorbing column[14] can only be responsible for an extinction $A_V \approx 0.45$ (assuming the Galactic A_V/N_H value[15]) and that A_V/N_H would need to be a factor 10–50 larger in order to block our line of sight to the broad line region in the optical. The existence of such a broad line region is inferred from the detection of broad Hα emission in the polarized light (Lumsden et al. 2001). Other broad (FWHM > 1000 km/s) permitted lines are not observed in this galaxy (Panessa & Bassani 2002). However, the direct spectrum also shows a broad component to the Hα line, i.e. it is a Seyfert 1.9 galaxy (Lumsden et al. 2001).

---
[14] The authors note that such a low column density would not necessarily imply obscuration in a torus, but a nuclear starburst or dust lanes on scales $\gg$ 1 pc would probably suffice. This is corroborated by the fact that most of the galaxies in their sample have large-scale bars or other dusty features that could provide the obscuration.

[15] From reddening in the NLR, Panessa & Bassani (2002) determined A_V and from X-Ray observations they found N_H. With these values they find no deviation from the Galactic value for A_V/N_H for their sample of unabsorbed Seyfert 2 galaxies. Besides, evidence for anomalous dust in AGNs points in the other direction than what would be needed to explain the observations of NGC 5995: the A_V/N_H value was found to be *lower* than the galactic value in X-Ray selected AGNs by Maiolino et al. (2001).

5.4. Radial visibility models

In this respect, it is remarkable that we see a clear decrease in correlated flux with increasing baseline length in the MIDI data, implying a resolved emitter of ≈ 3-4 pc in FWHM size. How can this be reconciled with the X-Ray classification as an unabsorbed Seyfert? Is NGC 5995 maybe simply a type 1 object with a weak broad line region, not requiring any obscuration in the optical at all? This would imply similar strengths for the broad polarized Hα and the broad component to the direct line. Unfortunately Lumsden et al. (2001) do not give a decomposition of the detected lines that would allow to test this hypothesis.

In single-dish mid-IR observations, Horst et al. (2009) find a slightly elongated nucleus at PA ≈ 110°. With the current dataset we cannot constrain any possible elongation as our observations so far only cover one position angle in (u, v) space.

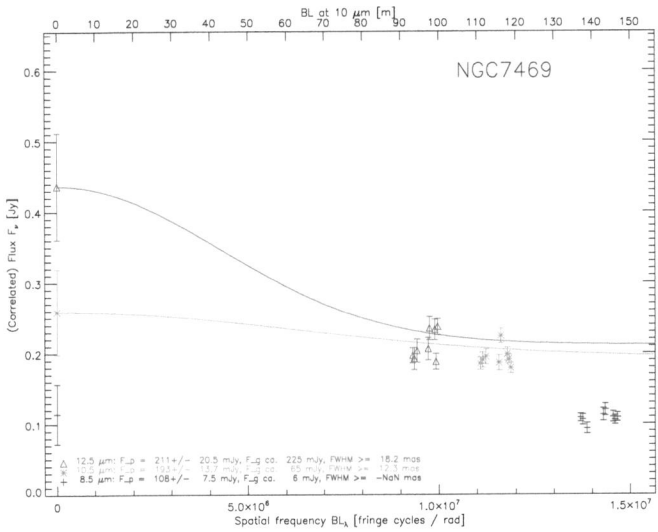

Figure 5.21.: Same as Fig. 5.7 but for **NGC 7469**

NGC 7469 This interacting Seyfert 1 galaxy is well known for its face-on starburst ring at a radius of ≈ 1″ (340 pc) from the nucleus (Díaz-Santos et al. 2007).

From its unresolved core in H band, Prieto et al. (2010) derive an upper limit to its FWHM size of 0.08″ (26 pc).

Soifer et al. (2003) observed NGC 7469 with the Keck I telescope in the mid-infrared and found a marginally resolved, elongated structure from the deconvolved image. They give a size of < 40 × 80 mas (< 13 × 26 pc) with the major axis of the structure at 135°.

In the GTO study the source was found to be well resolved (Tristram et al. 2009). For the LP analysis, the data from the GTO programme were not used, however, since it did not pass the aforementioned quality criteria (they have a very low number of good frames).

NGC 7469 is one of the weakest sources of the Large Programme. For this source we only have very few successful observations on the long U1U4 baseline at PA $\approx 45°$, i.e. roughly the PA of the minor axis of the elongated structure found by Soifer et al. (2003). At 12.5 μm, the correlated fluxes are clearly lower than the single-dish flux, showing an extended component of Gaussian FWHM $\gtrsim 18$ mas (5.4 pc). This limit is compatible with the upper limit of < 13 pc, given by Soifer et al. (2003), constraining the size of the mid-IR emitter in that direction to 5.4 pc $\lesssim$ FWHM $\lesssim 13$ pc. At 10.5 μm and 8.5 μm the source is compatible with being unresolved.

Mrk 463 E is a well-known interacting galaxy (for an image of its double nucleus see Mazzarella & Boroson 1993).

It was the only target selected for the Large Programme for which no observation was successful.[16]

---

[16]S. Hönig noted about the run in the excellent night of 26 March 2010 that the fringe search failed already on the short U1U2 baseline, even after solving an initial confusion which of the two nuclei was the eastern one. No further attempts were made to observe this source.

Table 5.2.: Fit results for the radial models, see text for explanations.

source	type	λ μm	F_p mJy	F_g mJy	Θ_g mas (pc)	χ_r^2 $\chi^2/\#DOF$
IZwicky1	Sy 1	8.5	170± 1.9	132± 19.4	30.6± 1.8 (32.9± 1.9)	146/16= 9.2
IZwicky1	Sy 1	10.5	268± 2.5	245± 26.0	33.9± 1.5 (36.5± 1.6)	177/16= 11.1
IZwicky1	Sy 1	12.5	286± 3.9	288± 32.6	39.2± 2.0 (42.2± 2.1)	81/16= 5.1
NGC1365	Sy 1	8.5	169± 1.9	97± 17.9	25.2± 1.8 (2.2± 0.2)	320/23= 13.9
NGC1365	Sy 1	10.5	227± 2.7	191± 24.2	28.7± 1.4 (2.5± 0.1)	483/23= 21.0
NGC1365	Sy 1	12.5	264± 4.4	181± 27.4	31.0± 2.2 (2.7± 0.2)	262/23= 11.4
LEDA17155	Sy 2	8.5	144± 5.3	68± 26.5	20.1± 4.5 (16.3± 3.6)	78/ 6= 13.1
LEDA17155	Sy 2	10.5	114± 7.5	123± 30.8	22.2± 4.0 (18.0± 3.3)	16/ 6= 2.8
LEDA17155	Sy 2	12.5	236± 9.2	182± 49.5	29.4± 4.5 (23.8± 3.6)	21/ 6= 3.5
MCG-5-23-16	Sy 2	8.5	105± 2.1	199± 20.9	21.3± 0.9 (4.0± 0.2)	456/20= 22.8
MCG-5-23-16	Sy 2	10.5	99± 12.0	211± 8.4	15.9± 1.4 (3.0± 0.3)	475/20= 23.8
MCG-5-23-16	Sy 2	12.5	148± 13.7	472± 19.3	21.1± 1.2 (4.0± 0.2)	494/20= 24.7
Mrk1239	Sy 1	8.5	326± 36.1	≈ 0	—	2.2
Mrk1239	Sy 1	10.5	445± 29.8	≈ 0	—	1.3
Mrk1239	Sy 1	12.5	450± 28.8	≈ 0	—	0.7
NGC3281	Sy 2	8.5	146± 9.4	≈ 166	≳ 28.4 (≳ 6.6)	
NGC3281	Sy 2	10.5	123± 19.2	≈ 91	≳ 25.9 (≳ 6.0)	
NGC3281	Sy 2	12.5	284± 28.3	≈ 347	≳ 39.0 (≳ 9.0)	
3C273	Quasar	8.5	196± 10.6	≈ 62	≳ 37.4 (≳ 98.9)	
3C273	Quasar	10.5	211± 16.3	≈ 62	≳ 40.1 (≳106.0)	
3C273	Quasar	12.5	268± 25.6	≈ 57	≳ 37.0 (≳ 98.1)	
NGC4507	Sy 2	8.5	58± 37.3	90± 20.4	12.2± 6.1 (3.0± 1.5)	224/ 5= 44.9
NGC4507	Sy 2	10.5	84± 55.7	167± 33.4	14.5± 5.5 (3.6± 1.4)	288/ 5= 57.8
NGC4507	Sy 2	12.5	111± 52.5	280± 26.3	19.9± 5.6 (5.0± 1.4)	30/ 5= 6.1
NGC4593	Sy 1	8.5	83± 62.2	32± 37.8	8.7± 20.9 (1.9± 4.5)	5/ 4= 1.5
NGC4593	Sy 1	10.5	130± 6.1	76± 43.1	18.9± 6.6 (4.0± 1.4)	33/ 4= 8.3
NGC4593	Sy 1	12.5	141± 8.8	103± 65.9	26.7± 7.3 (5.7± 1.6)	5/ 4= 1.4
IC4329A	Sy 1	8.5	413± 44.2	≈ 170	≳ 23.3 (≳ 7.7)	
IC4329A	Sy 1	10.5	561± 53.8	≈ 253	≳ 30.9 (≳ 10.2)	
IC4329A	Sy 1	12.5	624± 55.2	≈ 373	≳ 40.9 (≳ 13.5)	
NGC5506	Sy 2	8.5	398± 2.3	300± 43.5	19.8± 1.1 (2.8± 0.2)	627/ 6= 104.6
NGC5506	Sy 2	10.5	329± 2.5	300± 63.9	28.8± 1.8 (4.0± 0.3)	361/ 6= 60.2
NGC5506	Sy 2	12.5	561± 4.7	599± 65.9	27.0± 1.3 (3.8± 0.2)	319/ 6= 53.2
NGC5995	Sy 2	8.5	110± 3.8	150± 73.8	20.7± 3.6 (10.2± 1.8)	52/ 2= 26.0
NGC5995	Sy 2	10.5	117± 7.2	249± 96.6	22.8± 4.3 (11.3± 2.1)	25/ 2= 12.6
NGC5995	Sy 2	12.5	100± 226.6	228± 107.8	15.8± 19.8 (7.8± 9.8)	5/ 2= 2.5
NGC7469	Sy 1	8.5	108± 7.5	≈ 6	—	
NGC7469	Sy 1	10.5	193± 13.7	≈ 65	≳ 12.3 (≳ 3.6)	
NGC7469	Sy 1	12.5	211± 20.5	≈ 225	≳ 18.2 (≳ 5.4)	

5. The MIDI Large Programme: A statistical sample of resolved AGN tori

Table 5.3.: Fit results for the three other targets, compare to Table 5.2 and see text for explanations.

source	type	λ	F_p	F_g	Θ_g
		μm	mJy	mJy	mas (pc)
Circinus	Sy 2	10.5	570± 60	3070± 310	100± 10 (1.9± 0.2)
NGC1068	Sy 2	10.5	1700± 170	9500± 950	50± 10 (3.5± 0.4)
NGC4151	Sy 1	8.5	119± 16	410± 100	29± 5 (2.0± 0.3)
NGC4151	Sy 1	10.5	194± 61	700± 160	23± 4 (1.5± 0.3)
NGC4151	Sy 1	12.5	290± 70	1030± 300	32± 6 (2.1± 0.4)

5.4.1.1. Other targets

For the three other targets listed in Table 5.1, the results given in the literature were re-interpreted to be used in the context of the radial models.

The Circinus galaxy This galaxy has been studied in great detail, not only in the mid-infrared (see Tristram et al. 2007, and references therein). In the mid-IR, it is the second brightest southern galaxy and it was the second galaxy to be observed with MIDI. Its mid-IR emission is best fitted by a two component model consisting of a highly elongated disk and a large, round, component, according to Tristram et al. (2007). The elongated disk's major axis FWHM is 21 mas (0.4 pc) and the minor axis is unresolved. The large component was overresolved and contributes only marginally even at most of the shortest baselines. Its FWHM size is given as 100 mas (2 pc).

In their fit to the correlated and single-dish spectra, Tristram et al. (2007) determined *one* best-fit size of the emission (for each component) and did not fit a wavelength-dependent size like in our models. Since, in our models, the size of the emitter is almost always seen to be larger at 12.5 μm than at 8.5 μm , we interpret the fitted sizes of Tristram et al. (2007) as the average size that most likely corresponds to the size that we would give at 10.5 μm . To compare their results to the results from the 1-dimensional point source + Gaussian models, we further identified their disk-like component with the point source of our models and the larger component as the circularly symmetric Gaussian component.

The motivation for this is twofold: (1) We do not detect any deviations from circular symmetry in most of the LP sources. Any disk-like component (should it exist) must be unresolved. (2) If a disk component exists and does not scale with AGN luminosity, it would be unresolved in all of the LP targets due to their large distances.

The 10.5 μm fluxes (+ errors) of the disk component and the overresolved Gaussian component are taken from their Figure C.1 and are given in Table 5.3 together with the other relevant fit parameters. The error of the size of the large component is not given by Tristram et al. (2007) and therefore estimated to be $\approx$ 10 %.

NGC 1068 This mid-IR brightest extragalactic target in the southern sky was the first galaxy to be studied with MIDI (Jaffe et al. 2004). As for the Circinius galaxy, Raban et al. (2009) also fitted a two component structure consisting of a hot disk-like component of approximately 20 × 6.5 mas (1.35 × 0.45 pc) ("component 1") and a larger, less constrained, second component. This "component 2" is ca. 50 mas (3 × 4 pc) in Gaussian FWHM. The fluxes F_p and F_g are given in their Figure 2 and the errors of both the fluxes and the size of component 2 are estimated to be ≈ 10 %.

As in Circinus, their best model also gives one size for all wavelengths[17] and we interpret it as the 10.5 µm size. Also, their disk component will again be identified with the point source of our models. The resulting parameters are given in Table 5.3.

NGC 4151 The model applied for the mid-IR emission of NGC 4151 in Section 4 is very similar to the model a discussed here: In both cases the model consists of a point source and a resolved Gaussian source. However, since we have only 2 correlated flux observations for NGC 4151, the model has no free parameters and is therefore an exact solution (and not a fit). The flux errors for NGC 4151 are considerably larger (≈ 20%) than for other sources of this magnitude because this source had to be observed at very high airmass where an accurate calibration is very challenging. See Chapter 4 for further discussion.

...and Centaurus A? We do not include Centaurus A in the analysis here since this source is of a very different kind: Its mid-IR emission is very weak and at least half of it is non-thermal (Meisenheimer et al. 2007). The resolved emission in this source is probably a very different entity than the resolved mid-IR emission in Seyfert galaxies. See Chapter 3 for further discussion.

---

[17]Raban et al. (2009) also fit a model with wavelength-dependent size of the emitter. For consistency with the Circinus model, the wavelength-independent model is used for the discussion here.

5.5. Discussion

5.5.1. Torus scaling relations

One of the goals of the Large Programme is to look for statistical evidence of unification on the parsec scale. There, the obscuring dust is a central ingredient in unified models. Strict unification models that require it to be identical in all AGNs, were already excluded by the variety of dust seen in the first MIDI observations (see Section 1.4). Statistical unification models are less strict but still require the dust to appear differently in type 1 and type 2 galaxies. If such unified models contain any predictive power, the difference in appearance (e.g. the mid-IR size) of type 1 and type 2 tori must be dominated by class membership and not by individual properties of a galaxy.

Reverberation mapping studies (Suganuma et al. 2006; Landt et al. 2011) as well as near-IR interferometric observations (Kishimoto et al. 2007) suggest a relationship between the size of the innermost radius of hot dust and the illuminating luminosity, as predicted by theoretical models (Barvainis 1987).

It is expected that the radius of mid-IR emission also increases with illuminating radiation and that the study of the mid-IR size–luminosity relation will help to constrain torus models (Tristram & Schartmann 2011). Since the illuminating radiation, i.e. the UV / X-Ray flux, cannot be observed directly in type 2 objects, a relation between the mid-IR and the (absorption corrected) hard X-Ray luminosities of AGNs comes in handy (Krabbe et al. 2001; Horst et al. 2008) to relate observed to intrinsic properties. It implies that the mid-IR luminosity is a good tracer for the irradiating luminosity of the AGN.

Tristram et al. (2009) have shown a first size–luminosity relation for AGN tori (see their Figure 8) from the MIDI GTO sample of AGNs and confirmed the expected relation of $\Theta \propto \sqrt{\nu F_\nu}$. As expected from the diversity of dust distributions, the scatter in the relation is considerable. The GTO data are however incomplete in many respects: many data points are only limits, since the GTO observations did not provide the SNR and the (u, v) coverage to fit geometrical models to the visibilities. Only upper or lower limits were given where no correlated flux was observed or where no deviation from $V = 1$ was seen, respectively.

With the improved data set from the Large Programme we can now constrain a possible size–luminosity relation much better. Before turning to the results, however, let us briefly discuss which loci of such a size–luminosity diagram are actually observable with MIDI.

5.5.1.1. Observational constraints

Two obvious constraints limit the range of observability: (1) The total luminosity emitted by a blackbody (of constant temperature and emissivity) is proportional to its surface area (i.e. there are no small but luminous tori), (2) structures $\gg 1/BL_\lambda$ are over-resolved in an interferometer, leading to very low correlated fluxes. If there is no additional small component ("point source"), fringes cannot be tracked on such objects and they are not observable.

5.5. *Discussion*

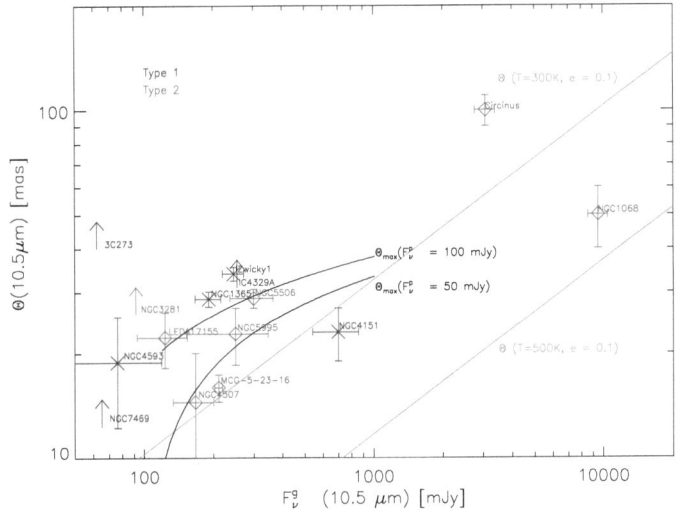

Figure 5.22.: Observability constraints for torus sizes with MIDI: Best-fit FWHM apparent size (Θ) as a function of resolved component flux (F_g). Type 1 objects are plotted as blue stars, type 2 objects are plotted as red diamonds. Arrows are plotted where only a lower limit to the FWHM size can be given. Overplotted are the expected loci of blackbodies of constant temperature and emissivity (green lines). Objects with a given point source flux F_p can only be observed up to a maximum size $\Theta_{\max}$ indicated by the black lines. NGC 1068 and the Circinus galaxies show "point sources" of $F_p > F_\nu^{\min}$, so that the $\Theta_{\max}$ criterion is not applicable to them.

In Figure 5.22 we plot, logarithmically, the best-fit FWHM size Θ as a function of the best-fit flux of the structure of this size, F_g. Overplotted are the observability constraints that are discussed below. In this plot, only *observed* quantities are plotted to clearly separate the observational limitations from any astrophysical effects.

(1) The loci of constant temperature blackbodies A blackbody emitter with constant temperature $T \approx 300$ K, emissivity (covering factor) $\epsilon \approx 0.1$ and full-width half maximum size Θ produces a (total) flux F of

5. The MIDI Large Programme: A statistical sample of resolved AGN tori

$$F \approx \nu F_\nu = \epsilon \sigma_{\text{SB}} \Theta^2 T^4 / 4 \qquad (5.8)$$

with the Stefan-Boltzmann constant $\sigma_{\text{SB}} = 5.67 \cdot 10^{-8} \text{Js}^{-1}\text{m}^{-2}\text{K}^{-4}$. The assumption $F \approx \nu F_\nu$ implies that the torus emits most of its flux at the observed wavelength. Rearranging the formula, a blackbody with an observed flux (density) F_ν has a Gaussian FWHM Θ of

$$\Theta = \sqrt{\frac{4\nu F_\nu}{\epsilon \sigma_{\text{SB}} T^4}} \qquad (5.9)$$

These are the straight green lines in Figure 5.22. It connects tori of constant temperature and emissivity.

(2) The flux limit For a given sensitivity limit $F_\nu^{\text{min}} \approx 150$ mJy (at 10.5 μm), observable targets must obey the relation

$$F_p + F_g \cdot \exp\left(-\frac{(\pi \Theta B L_\lambda)^2}{4 \ln 2}\right) \gtrsim F_\nu^{\text{min}} \qquad (5.10)$$

This translates into a maximum size of the resolved emitter

$$\Theta_{\text{max}} = \frac{2\sqrt{\ln 2 \ln(F_g/(F_\nu^{\text{min}} - F_p))}}{\pi B L_{\lambda,\text{min}}} \qquad (5.11)$$

with the minimum spatial resolution of the observations $BL_{\lambda,\text{min}} \approx 50\text{m} / 10.5$ μm $\approx 5 \cdot 10^6$. For $F_p \geq F_\nu^{\text{min}}$, Θ_{max} does not exist, i.e. tori of any size are observable. Whether any change of correlated flux with baseline length can be observed depends on the SNR of the respective observations.

At a given F_g, the maximum size an object may have to still be observable is given by the point source flux: Objects with little point source contributions must have smaller sizes (leading to larger correlated fluxes) to be observable. These maximum sizes are plotted in Figure 5.22 for point source fluxes of 50 mJy and 100 mJy.

From this Figure it can be seen that all of the resolved objects (except NGC 1068) are larger than expected for a 300K blackbody with emissivity of 10%, i.e. the observed structures trace emitters with lower surface brightnesses than such a blackbody.

There appears to be an indication that type 1 objects trace larger structures than type 2 objects – contrary to the expectation (Tristram & Schartmann 2011). However, one must be careful when interpreting this bias: Since type 1 sources typically have larger point source fluxes than type 2 sources (see also Figure 5.24 below), observational constraints limit our ability to detect large tori in type 2 sources. Type 2 sources with very large tori are simply not observable.

5.5. Discussion

The size–luminosity relation In Figure 5.23, the size of the resolved emitter at 10.5 μm is plotted as a function of luminosity $L = \nu L_\nu$, estimated at 12.5 μm. The reason for choosing two different wavelengths here is that we only know the sizes for *all* sources at 10.5 μm (see subsection 5.4.1.1), but the flux is better estimated well outside any possible silicate feature, e.g. at 12.5 μm.

Overplotted is the relation $s \propto L^{0.5}$. The offset has been set to 0 and the slope has not been fitted. It matches the data points remarkably well, though. What does the relation tell us? Does it imply that tori scale with luminosity?

First of all, a size–luminosity relation of $s \propto L^{0.5}$ is the trivial outcome for blackbody emission (see Equation 5.9). The fact that the data points show relatively little scatter around that relation indeed implies that the observed tori all have similar surface brightnesses (i.e. the product of covering factor × T^4).

As a future step, the observability constraints must be studied and discussed for such a diagram to better separate observability constraints from astrophysics.

5.5.1.2. Does distance matter?

In order to learn more about 'tori' (e.g. by studying their substructure, the 'clumpiness'), it is important to understand how the characteristic scale of their constituents ρ scales with source luminosity and distance r from the nucleus. While most toy models find or assume that $\rho \propto r^\beta$ with $\beta \approx 1$ (e.g. Hönig et al. 2006; Schartmann et al. 2008), this is not found in the hydrodynamical torus model of Schartmann et al. (2009) where the thickness of the filaments, which the torus is made of, is independent of distance. In this case, a fixed characteristic scale of the torus substructure, or equivalently assuming that the torus geometry simply depends on the observed flux, is no longer correct and rather a fixed spatial resolution would be needed. This would imply that we need to observe the closest AGNs in order to resolve the torus substructure (clumpiness).

Does Figure 5.23 already imply that tori scale with luminosity? Let us assume that tori are indeed made up of a relatively small disk component (as seen in NGC 1068 and Circinus) of size $\approx 0.1 pc/\sqrt{L/L_\text{Circinus}}$ and a larger "torus" component of $\approx 1 pc/\sqrt{L/L_\text{Circinus}}$. In that case – both components scale with luminosity – the "point source fraction" should be constant with distance, using the same arguments as before.

Now let us look at a plot of point source fraction f_p as a function of resolved scale (Figure 5.24). Two observations strike the eye:

1. As expected, type 1 galaxies have on average larger point source fractions than type 2s. This should be a relatively robust result since the observational bias explained above is inverse: it biases against type 2 galaxies with very low point source fractions.

2. Contrary to the assumption made above, it is also evident that the point source fraction is a function of distance: The larger the scale, the more point source contribution we see, indicating that the sources *do not scale* strictly scale as $L^{0.5}$ and the sizes are therefore not distance independent.

5. The MIDI Large Programme: A statistical sample of resolved AGN tori

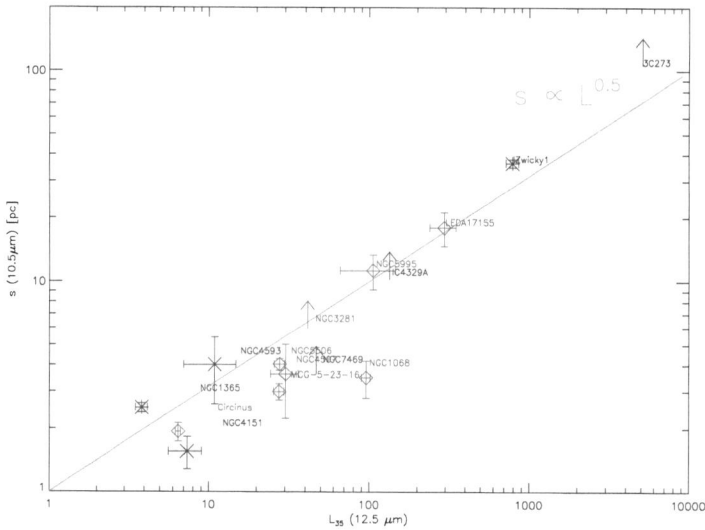

Figure 5.23.: Size–luminosity relation for the LP sample. The size s is given by $D_A \cdot \Theta_g$ where D_A is the angular-size distance, the luminosity is estimated from $L = 4\pi D_L^2 c/\lambda F_\nu$ at 12.5 μm, where D_L is the luminosity distance, $L_{35} = L/(10^{35}\text{W})$. The green line is $s = L^{0.5}$, i.e. the expected relation $s \propto L^{0.5}$ with zero offset.

The observation that the point source fraction increases with luminosity (point 2), can be understood in the context of the so called "receding torus paradigm" (Lawrence 1991). It predicts that the inner radius of the torus r_{in} increases with the AGN's luminosity L_{UV} as $r_{in} = 1.3 L_{UV,46}^{1/2} T_{1500}^{-2.8}$ pc (Barvainis 1987) – this follows from the fact that r_{in} is given by the sublimation temperature of the dust. Larger values of r_{in} lead to smaller torus half-opening angles and so, in the unified picture, the fraction of type 1 AGNs increases with luminosity. This is roughly compatible with source counts (Lawrence 1991). However, the exact form of the luminosity dependence of the type 1 fraction is inconsistent with the most simple receding torus paradigm in which the torus height h remains constant with luminosity (Simpson 2005).

In Figure 5.24, the scale axis is more or less identical with luminosity (for the LP targets) since they all have approximately the same observed flux.

5.5. Discussion

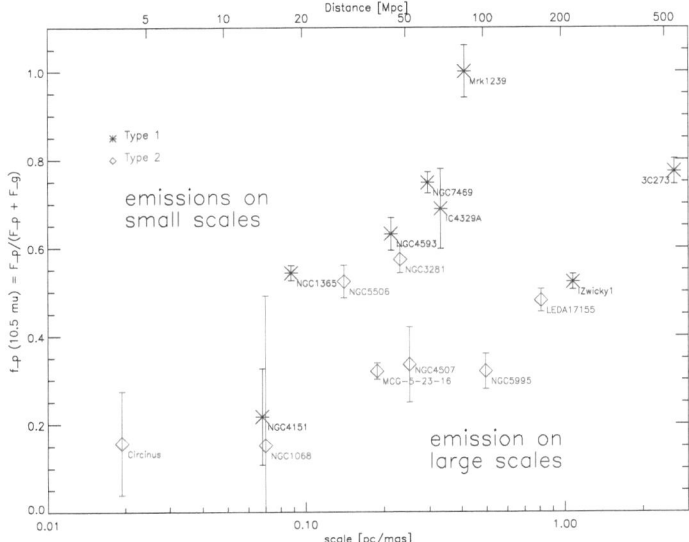

Figure 5.24.: Point source fraction as a function of resolved scale (distance). Type 1 objects have a larger point source fraction than type 2 objects. The more distant the source, the larger the point source fraction.

In the receding torus paradigm, the extinction at 10 μm to the inner parts of the torus would decrease (in the statistical average), i.e. the unresolved part gets brighter, offering a possible explanation for the increase in point source fraction.

5.5.2. The sub-structure of tori

5.5.2.1. Observational signs of torus substructure

A possible way to detect torus substructure was devised by K. Tristram: In a source where the baseline length does not change with position angle, one can study the source at the same spatial resolution for various position angles. For observations at the VLTI, this is fulfilled for objects at $\delta \approx -65°$, such as the Circinus galaxy. One can then compare these observations to fluxes from radiative transfer models to constrain torus parameters such as the characteristic size of the clumps. First results from such observations were

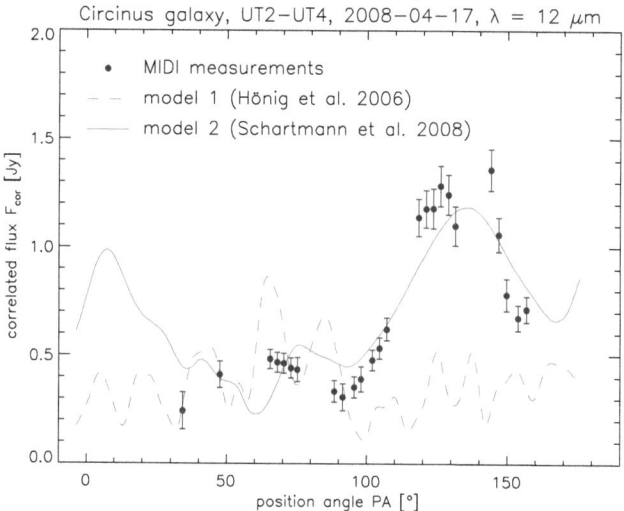

Figure 5.25.: Clumpiness in Circinus. The correlated flux increased about three-fold with a change of position angle of just about 30°. This cannot be explained by a simple elongation (inclination) of a disk since the latter would only produce variations with a period of 180°.
The two radiative transfer model curves differ mainly by the size of the clumps where the clumps in the model by (Schartmann et al. 2008) were much larger than the clumps in the model employed by Hönig et al. (2006). Figure courtesy K. Tristram

promising (Figure 5.25).

5.5.2.2. "Continuous fringe tracks" in the Large Programme

Most of the radial model are poor fits to the LP data in the sense that $\chi_r^2 \gg 1$. Have we under-estimated the errors? Probably not! Since we found no evidence for large systematic errors in the correlated fluxes[18], we argued that the (relatively robust) statistical errors are an adequate description of the measurement uncertainties.

In other words, the models are too simple and the remaining scatter tells us that there is more structure in the sources that we did not fit by using the smooth Gaussian "envelope" function.

Supporting evidence comes from the desired signs for clumpiness. To detect clumpy tori, some sources have been sampled very densely on baselines where the projected baseline

---

[18]The single-dish flux does not contribute very much to χ_r^2 due to its relatively large error.

5.5. Discussion

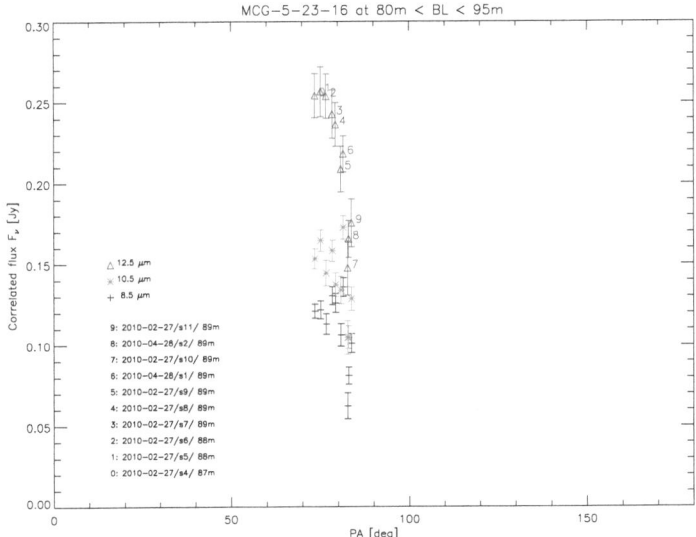

Figure 5.26.: Evidence for clumpiness in MCG-5-23-16? The correlated flux changes by almost a factor of two within less than 10° of change in position angle. Such a change can be interpreted in the context of clumpy models (see Figure 5.25).

is not a strong function of position angle (e.g. MCG-5-23-16, NGC 1365). These two sources already showed evidence of non-centrosymmetric structures from Figures 5.9 and 5.11 respectively. Let us now look at plots of correlated flux as a function of *position angle*, at a certain (narrow) range of projected baseline.

MCG-5-23-16 has a well sampled (u, v) plane (Figure 5.6) and is the best case for small-scale structure (clumpiness) in the Large Programme (Figure 5.26). Here, the correlated flux seems to be a function of position angle, decreasing from $\approx (250 \pm 15)$ mJy to $\approx (160 \pm 15)$ mJy, or by a factor of 0.6 over only $\approx 10°$ in projected baseline angle.

In this source, not only is the variation with position angle significant, but the observations also show a continuous decrease of correlated flux, *even though* the points marked as '6' and '8' have been taken in different nights than the others – corroborating the

5. The MIDI Large Programme: A statistical sample of resolved AGN tori

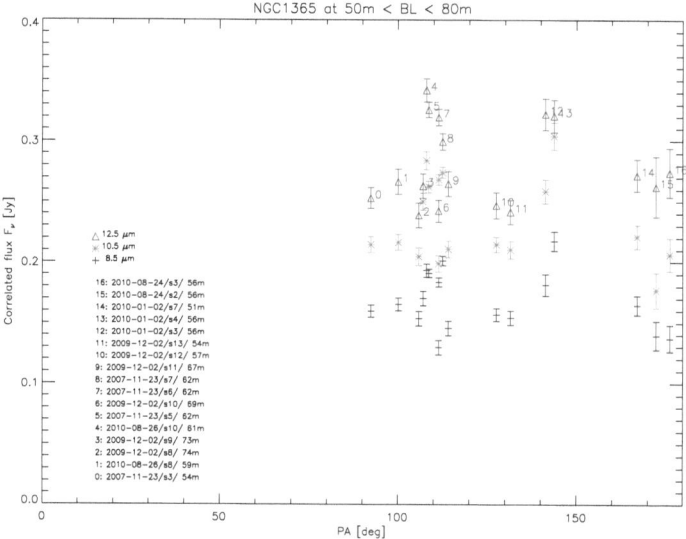

Figure 5.27.: Evidence for clumpiness in NGC 1365? The outlier point (6) does not fit into the clumpiness picture. A more detailed statistical investigation is necessary to evaluate the significance of this outlier.

argument that systematic errors do not dominate over statistical ones.

NGC 1365 also has a region of densely sampled (u, v) space at around $50 < \text{BL} < 80\text{m}$ and $100° < \text{PA} < 120°$ (Figure 5.6). If plotted against position angle, a decrease in flux can be observed, starting from the point labeled '4' to '9' and further to '10' and '11' (Figure 5.27). However, point '6' in between does not fit to this trend.

According to our selection criteria, there is no reason to believe that point '6' should be excluded. This point is at a slightly larger baseline (about 1 telescope diameter larger than neighboring points 5 and 7). This can, however, not be the reason for it being a significant outlier (see Figure 5.9).

5.6. Conclusions

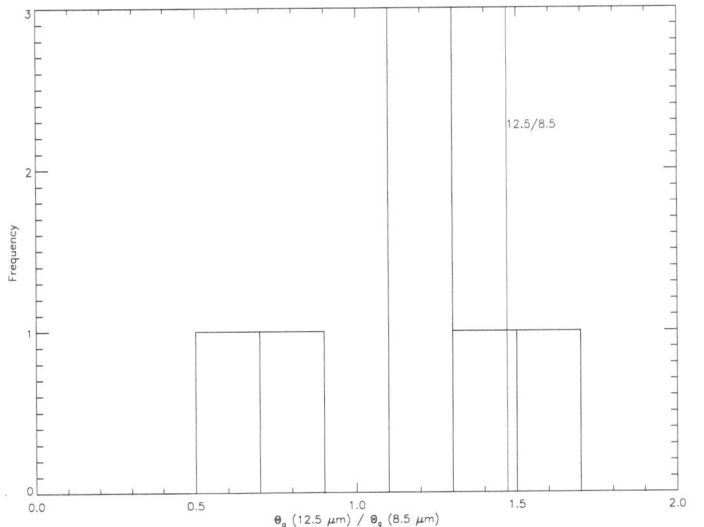

Figure 5.28.: Histogram of the ratio of torus size at 12.5 μm to the size at 8.5 μm. A value close to $12.5/8.5 \approx 1.5$ is expected.

5.5.2.3. Torus size as a function of wavelength

The histogram of torus size relations (Figure 5.28) shows that the size increase is roughly compatible with dust being heated from inside. Assuming greybody models where $T \propto r^{-\alpha}$ with $\alpha \approx 1/2.8$ (Barvainis 1987), the source size increases roughly proportional to λ since colder parts of the structure (that have their emission peak at longer wavelengths) are larger.

5.6. Conclusions

- The VLTI/MIDI AGN Large Programme is the largest study of resolved AGN nuclear dust so far.

- A large amount of data has been selected and reduced and new, more robust, ways to determine errors in MIDI observations have been developed.

5. The MIDI Large Programme: A statistical sample of resolved AGN tori

- The LP has been successful: most targets could be observed and showed signs for being resolved (albeit some only marginally). It is remarkable that correlated fluxes can now routinely be observed at the < 200 mJy level with MIDI!

- However, the range in visibilities is small and no elongation is seen in the observed sources.

- For the first fits we therefore applied only 1D models to constrain the size and flux of the emitter and the flux of the unresolved source.

- A plot of torus size s versus 12 μm luminosity L reveals a striking relation very close to the expected $s \propto L^{0.5}$ relation (for centrally heated dust). This could be an indication that tori experience similar heating mechanisms over more than three orders of magnitude in luminosity.

- It appears to be promising to look for differences between type 1 and type 2 sources in such size–luminosity relations. However, statistical studies from MIDI data suffer biases due to the many observability constraints. They must be studied carefully and understood before claiming statistical relations. Near the flux limit of MIDI, large type 2 sources are harder to detect than large type 1 sources since the latter almost always have larger point source contributions, making it easier to track fringes on them.

- A relation of point source flux contribution versus resolved scale shows that the point source contribution increases with increasing distance, i.e. with a larger resolved scale. Since most of the sources studied here are very similar in flux, distance is effectively proportional to luminosity. The fact that the point source contribution is not independent of distance (luminosity) implies that at least not all components of an AGN torus (i.e., in our models, the resolved source and the point source) scale with luminosity as $s \propto L^{0.5}$ (otherwise the point source fraction would be constant with distance).

- It was to be expected that a smooth visibility function, such as the Gaussians that we fit, cannot explain the complicated structure expected for AGN tori (e.g. from hydrodynamical modelling, Figure 1.6). It is nevertheless striking that one convincing case for resolved torus substructure was found, in the observations of MCG-5-23-16 where the correlated flux changes by about a factor of two within less than $10°$ of change in position angle, similar to what has been seen in the Circinus galaxy and expected in clumpy torus models.

5.7. Outlook

- A completion study is underway (087.B-0266, PI: K. Meisenheimer) to fill the gaps in the (u, v) planes caused by weather loss during the LP (see Figure 5.29).

5.7. Outlook

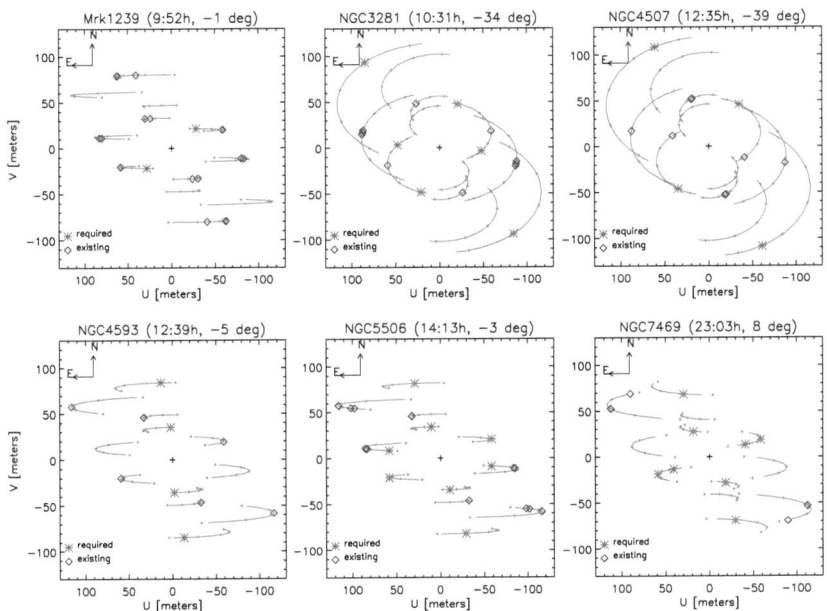

Figure 5.29.: The LP completion study (087.B-0266) is already scheduled and will help to fill the gaps in the (u,v) coverage caused by bad weather during the LP.

- VISIR data has been taken (P086.B-0919, PI: K. Tristram) of those sources where no good high-resolution single-dish spectra existed. This will lead to reduced uncertainties in the '0-baseline' points and help to constrain the models.

- Supporting programmes have been started that are connected to the LP, e.g. a SINFONI programme (PI: M. Schartmann) with the objective to test the link between star formation and AGN tori that is being explored successfully in hydrodynamical models.

- The selection criteria for the LP were obviously still very conservative: Since 12 of our 13 targets could be observed, we have probably not yet reached the limit of MIDI's sensitivity with this sample and even weaker sources seem to be observable.

- In the near future, we need more short baselines (and more sensitivity to do that with the ATs) to better constrain the relatively large sizes of the resolved emission found in many sources.

- In the more distant future, it would be desirable to observe these targets with even

5. The MIDI Large Programme: A statistical sample of resolved AGN tori

longer baselines to resolve the "point source" and see if it, in type 2 objects, really is a disk with a large axis ratio as seen in the nearby galaxies NGC 1068 and Circinus.

6. Conclusions

Methodical

Mid-infrared interferometry is a very powerful tool for resolving structures 10 times smaller than the spatial resolution of even the largest single-dish telescopes available today. However, it is also a complex observing method involving many highly sophisticated technical subsystems to overcome the huge background in the thermal infrared. MIDI at the VLTI is currently the most sensitive interferometric instrument for long-baseline interferometry in the mid-IR, now routinely observing targets with correlated fluxes as low as $F_{\nu,\mathrm{corr,lim}}(12\mu\mathrm{m}) \approx 200$ mJy. Within the course of this thesis, we have developed both the observing method and the data reduction and calibration technique in order to gather acceptable data for such sources as well as reliably determine the error in the calibrated fluxes.

From the re-observation of identical (u, v) points, confidence was gained that the flux of weak sources can actually be determined to better than 10 % under good conditions. An analysis of atmospheric fluctuations further indicated that, for good nights, the transfer function variations do not prohibit the direct calibration of target correlated fluxes with calibrator correlated fluxes, bypassing the very noisy per-observation visibility estimation (i.e. dividing each correlated flux by the subsequently taken very noisy single dish flux) that is the standard observing technique for brighter sources.

Astrophysical

In a multi-baseline campaign, the (u, v) coverage of the nearest radio galaxy and major merger, Cen A, was substantially extended compared to previous observations. The resulting visibility pattern defies any easy explanation. After a careful study of the statistical and systematic errors, we are convinced that the pattern is real and that the bad fits to simple smooth models of an elongated Gaussian emitter and an unresolved point-source actually indicate a more complex structure. A better fit is indeed found if an offset of ≈ 40 mas (0.7 pc) at a $PA \approx 45°$ is introduced between the unresolved point source and the extended source. The flux of the unresolved source is found to contribute roughly 55 % (1.00 ± 0.24 Jy, at 12.5 ± 0.2 μm) to the total flux with the rest (0.82 ± 0.2 Jy) emitted by an elongated structure. Its size is unresolved along the minor axis. The major axis is at a $PA = 115.5 \pm 1.5°$ and can be estimated to be ≈ 48 mas at 12.5 μm in size (FWHM of the Gaussian model component). Furthermore, a flux increase by about 50 % is detected in the mid-IR between 2005 and 2008 and an outburst is discovered in X-Ray monitoring data. Using this constraint, a scenario is constructed that allows us to locate

6. Conclusions

the position of the resolved emitter very near to the front-facing part of the nuclear jet suggesting a physical connection between the offset mid-infrared emitter and the jet.

In challenging observations at extremely high airmass, the brightest type 1 galaxy NGC 4151 was successfully observed and a resolved thermal emitter is found as well as tentative evidence for silicate emission on the parsec scale. The properties of the resolved emitter (Gaussian FWHM size 2.0 ± 0.4 pc, color temperature 285^{+25}_{-50} K, emissivity $\approx 10\%$, temperature gradient $T \propto r^{-0.36}$) are found to be similar to the tori that have been resolved in type 2 objects before, showing consistency with but not necessarily confirming line-of-sight unification schemes. Recent, statistical unification models (e.g. Elitzur & Shlosman 2006) predict a wide variety of nuclear dust structures in any AGN type and can therefore only be studied on a statistical basis.

This is the goal of the Large Programme, an observational campaign including 13 AGNs of various types and luminosities. More than twice as much data than during the whole MIDI GTO AGN programme were collected, reduced and calibrated within less than a year. In all but one source, resolved emitters are found ranging in Gaussian FWHM size from about 1 pc to $\gtrsim 100$ pc (at 10.5 μm), making this the largest sample of resolved AGN tori to date. Many of the sizes can be determined to an accuracy of $\approx 10 - 20\%$, impressively demonstrating the improvements made both at the VLTI and for the data reduction software, since almost all of these sources have correlated fluxes that would probably have been unobservable a few years ago.

The sizes s of these structures seem to scale with luminosity νL_ν as $s \propto L^{0.5}$, as expected for centrally illuminated dust – but also just for any blackbody of constant emissivity and temperature – suggesting that the tori studied in this data set are comparable in surface brightness.

In the search for correlations, a number of observational constraints must be taken into account in order to deduce astrophysical results. For example, large structures are harder to detect in type 2 sources than in type 1 sources since the former have lower point source contributions and therefore their correlated fluxes may drop below $F_{\nu,\text{corr,lim}}$ for highly resolved sources. In the most simple torus scaling relations the torus size would scale with luminosity and thus with distance since our sample is comprised of nearly equally bright galaxies. The fact that the point source flux contribution seems to not only depend on class membership but also on the distance of the source, suggests that such simple scaling relations do not hold.

In summary, I have spent a good part of my doctoral studies with understanding, applying and maybe even developing the technique of mid-IR interferometry, a wonderfully elegant but challenging observational method. At the same time, the motivation for the studies of resolved nuclear dust has clearly evolved from being dominated by unification questions into more physical questions concerning the nature of AGN tori.

7. Outlook

7.1. Developments at the VLTI

Sensitivity and efficiency increase for MIDI at the VLTI Current efforts to further increase both the sensitivity and the efficiency for MIDI observations at the VLTI include

- a planned technical study for "hybrid" observations, combining ATs and UTs. They would offer an increase in correlated flux sensitivity by $\approx \sqrt{F_{\mathrm{UT}}/F_{\mathrm{AT}}} \approx 4$ with respect to AT-AT combinations and also provide short baselines and new position angles, not available with the six UT baselines.

- a recently added new mode ("correlated flux mode") offers to skip the noisy single-dish observations for weak science sources *also in service mode* (Rivinius, Th 2011). This allows more flexibility for the planning of observations. A further improvement called "remember-go-back" mode is being discussed, allowing to store the MACAO filter settings for both a science and a calibrator source, thereby reducing the large overheads due to the search of the right MACAO filter (see Section 2.5). This would allow a very dense sampling of the (u, v) plane with up to 3 target and calibrator correlated fluxes per hour.

- A further improvement will be reached, once PRIMA (see Section 2.5), will be offered as a fringe tracker for MIDI. It has been demonstrated that it will be possible to track fringes *with the ATs* for sources as faint as $F_{\nu,\mathrm{corr}} \approx 500$ mJy using the PRIMA fringe tracker in combination with MIDI (Müller et al. 2010a). The instrument is currently under comissioning.

Imaging interferometry The next big step for optical long-baseline interferometry in the mid-infrared will be the beam combination from more than two telescopes. This way a relation for the visibility phase between the three telescopes (the so called closure phase) can be deduced, providing the ability to produce actual images of objects. Apart from that, the combination of more than two telescopes increases the instantaneous (u, v) coverage. With 4 telescopes, for example, 6 (u, v) points will be observed simultaneously making it possible to obtain well sampled (u, v) planes in shorter time than with a two telescope combiner.

Currently, two 4-beam combiners are being developed for the VLTI: MATISSE (for the mid-IR) and GRAVITY (in the near-IR). In the context of AGN torus studies, MATISSE is of particular relevance since the *images* that it will produce (see Figure 7.1) will offer a more direct way of comparing AGN models with observations.

7. Outlook

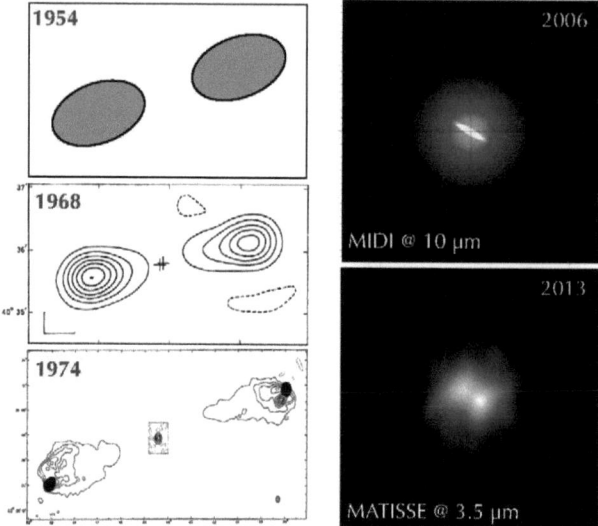

Figure 7.1.: Comparison of the development of radio images by the example of the radio galaxy Cygnus A with the expected development of mid-IR interferometric results by the example of the Circinus galaxy (Meisenheimer 2008).

7.2. ALMA and AGNs

The Atacama Large Millimeter Array (ALMA), an interferometer currently being built, will offer exciting possibilites to study AGN tori (e.g. Maiolino 2008). With a spatial resolution in the sub-mm comparable to the VLTI's resolution at $10\mu m$, the combination of the two will offer an even broader view towards this complex region: by observing the nuclear region with milli-arcsecond resolution over more than two orders of magnitude in wavelength (10 μm – 3 mm), constraints for the structure of tori (smooth / clumpy distributions) can be derived. By spatially resolving emission lines of individual clumps of gas and dust with ALMA, one can look for outflow/inflow/rotational/turbulent motions to constrain the formation history and dynamical state of the torus (Large-scale outflows? Supernova-induced turbulence? Orderly rotation or inflow?).

Acknowledgements

First and foremost, I wish to thank my *Doktorvater* Klaus Meisenheimer for accepting me as his doctoral student and for extraordinary support during the entire thesis. His professional and personal advice was essential for me and his support was especially helpful when apparently unsurmountable problems were to be solved.

Discussions with Walter Jaffe (Leiden) were elemental for my understanding of interferometry. Walter always took the time to answer my questions in great detail – and extended his hospitality to me during my visit to Leiden in October 2008. This visit was supported by an EU/Opticon Fizeau stipend.

I wish to thank my referees, Klaus Meisenheimer and Sebastian Wolf for reviewing my thesis and my thesis committee members Max Camenzind, Klaus Meisenheimer and Sebastian Wolf for their advice.

My colleague Konrad Tristram (Bonn) helped me to quickly gain experience with interferometric data reduction and my office mate Roy van Boekel was an invaluable partner for discussions about pretty much everything.

I would also like to thank René Andrae, Gaëlle Dumas, Uwe Graser, Knud Jahnke, Christoph Leinert, Christian Leipski, Carole Mundell, Jörg-Uwe Pott, Hans-Walter Rix, Marc Schartmann, Andreas Schruba, Helmut Steinle and Steven Tingay for helpful discussions.

I am thankful to the MPIA staff for making this such an enjoyable place to work and also to ESO astronomers and other staff both at the VLT and in the guesthouse in Santiago, Chile, for their professional support and hospitality.

I would like to thank Myriam Benisty, Henrik Beuther, Roy van Boekel, Helmut Burtscher, Helmut Dannerbauer, Christian Fendt, Mathias Jäger, Knud Jahnke, Christian Leipski, Klaus Meisenheimer, André Müller and Laura Watkins for proof reading and comments on my thesis.

For financial support during my thesis, I would like to acknowledge the Max Planck Society and the International Max-Planck Research School for Astronomy and Cosmic Physics (IMPRS-HD).

This work would not have been possible without the support I receive from my family. I wish to thank my mother, Irmgard Burtscher, and my mother-in-law, Sigrun Overesch, for great support especially in the last months of writing. Most of all, though, I am indebted to my wife, Cordula Burtscher for her loving care.

This research is based on observations collected at the European Organization for Astronomical Research in the Southern Hemisphere, Chile, programme numbers 081.B-0092(A) (NGC 4151), 081.B-0121* (Centaurus A) and 184.B-0832 (LP) and has made use of

- data obtained from the ESO Science Archive Facility under various request numbers.
- the Jean-Marie Mariotti Center LITpro service co-developped by CRAL, LAOG and FIZEAU,
- quick-look results provided by the ASM/RXTE team,
- NASA's Astrophysics Data System,
- the NASA/IPAC Infrared Science Archive, which is operated by the Jet Propulsion Laboratory, California Institute of Technology, under contract with the National Aeronautics and Space Administration and

A. List of Abbreviations

ADAF Advection-Dominated Accretion Flow

AGN Active Galactic Nucleus

AO Adaptive Optics

AT Auxiliary Telescope

BH Black Hole

BLR Broad Line Region

CCW Counter clockwise

CW Clockwise

DIMM DIfferential Motion Monitor (small telescope that measures the seeing on Paranal)

DIT Detector Integration Time

DDT Director's Discretionary Time – an ESO category for observing programmes that require little observing time but might have great impact. DDT proposals can be submitted at any time.

DOF Number of degrees of freedom (in a χ^2 fit)

EWS Expert Work Station, interferometric data reduction package by W. Jaffe and others

FWHM Full width at half maximum

GTO Guaranteed Time Observations – time rewarded to an institute or consortium for building an instrument

HA Hour Angle

HWHM Half width at half maximum

IRIS Infrared Image Sensor

JMMC Jean-Marie Mariotti Center

Jy Jansky, 1 Jy = $10^{-26}\,\mathrm{W\,m^{-2}\,Hz^{-1}}$

A. List of Abbreviations

LINER Low-Ionization Nuclear Emission Region

LITpro Lyon Interferometric Tool prototype, an interferometric model fitting software by JMMC

LP Large Programme – an ESO category for observing programmes requiring more than 100 hours of observing time

MACAO Multi-Application Curvature Adaptive Optics, the Coudé Adaptive Optics system on the VLT UTs

mas milli-arcsecond(s), $1/1000'' = 4.848 \cdot 10^{-9}$ rad

MIDI MID-infrared Interferometric instrument

NLR Narrow-Line Region

OB Observing Block

OPC Observing Programmes Committee, (ESO) panel that decides about observing proposals

OPD Optical Path Difference, also called delay

PSF Point Spread Function

PWV Precipitable Water Vapor, the water column in the atmosphere

rad radian

SMBH Super-Massive Black Hole

SNR Signal to Noise Ratio

TIO Telescope and Instrument Operator

UT Unit Telescope

VLT(I) Very Large Telescope (Interferometer)

ZOPD Zero Optical Path Difference

Bibliography

Agudo, I., Gómez, J., Martí, J., Ibáñez, J., Marscher, A. P., Alberdi, A., Aloy, M., & Hardee, P. E. 2001, ApJ, 549, L183

Antonucci, R. R. J., & Miller, J. S. 1985, ApJ, 297, 621

Armstrong, J. T., et al. 1998, ApJ, 496, 550

Arsenault, R., et al. 2003, in Presented at the Society of Photo-Optical Instrumentation Engineers (SPIE) Conference, Vol. 4839, Society of Photo-Optical Instrumentation Engineers (SPIE) Conference Series, ed. P. L. Wizinowich & D. Bonaccini, 174–185

Bakker, E. J., et al. 2003, in Presented at the Society of Photo-Optical Instrumentation Engineers (SPIE) Conference, Vol. 4838, Society of Photo-Optical Instrumentation Engineers (SPIE) Conference Series, ed. W. A. Traub, 905–916

Baldwin, J. A., Phillips, M. M., & Terlevich, R. 1981, PASP, 93, 5

Baldwin, J. E., Haniff, C. A., Mackay, C. D., & Warner, P. J. 1986, Nature, 320, 595

Barlow, R. 1989, Statistics. A guide to the use of statistical methods in the physical sciences (The Manchester Physics Series, New York: Wiley, 1989)

Barvainis, R. 1987, ApJ, 320, 537

Beckert, T., Driebe, T., Hönig, S. F., & Weigelt, G. 2008, A&A, 486, L17

Begelman, M. C., Blandford, R. D., & Rees, M. J. 1984, Reviews of Modern Physics, 56, 255

Bekenstein, J. 2006, Contemporary Physics, 47, 387

Bentz, M. C., et al. 2006, ApJ, 651, 775

—. 2008, ApJ, 689, L21

Berger, J. P., & Segransan, D. 2007, New Astronomy Review, 51, 576

Best, P. N. 2009, Astronomische Nachrichten, 330, 184

Bibliography

Bester, M., Danchi, W. C., & Townes, C. H. 1990, in Presented at the Society of Photo-Optical Instrumentation Engineers (SPIE) Conference, Vol. 1237, Society of Photo-Optical Instrumentation Engineers (SPIE) Conference Series, ed. J. B. Breckinridge, 40–48

Bland, J., Taylor, K., & Atherton, P. D. 1987, MNRAS, 228, 595

Blandford, R. D., & McKee, C. F. 1982, ApJ, 255, 419

Bolton, J. G., Stanley, G. J., & Slee, O. B. 1949, Nature, 164, 101

Braatz, J. A., Wilson, A. S., & Henkel, C. 1994, ApJ, 437, L99

Brown, R. H., & Twiss, R. Q. 1957, Royal Society of London Proceedings Series A, 242, 300

Bruhweiler, F., & Verner, E. 2008, ApJ, 675, 83

Buchanan, C. L., Gallimore, J. F., O'Dea, C. P., Baum, S. A., Axon, D. J., Robinson, A., Elitzur, M., & Elvis, M. 2006, AJ, 132, 401

Burtscher, L. 2011, PhD thesis, Naturwissenschaftlich-Mathematische Gesamtfakultät der Ruprecht-Karls-Universität Heidelberg, electronic version available at http://www.ub.uni-heidelberg.de/archiv/12037/

Burtscher, L., Jaffe, W., Raban, D., Meisenheimer, K., Tristram, K. R. W., & Röttgering, H. 2009, ApJ, 705, L53

Burtscher, L., Meisenheimer, K., Jaffe, W., Tristram, K. R. W., & Röttgering, H. J. A. 2010, PASA, 27, 490

Capetti, A., et al. 2000, ApJ, 544, 269

Cappellari, M., Neumayer, N., Reunanen, J., van der Werf, P. P., de Zeeuw, P. T., & Rix, H.-W. 2009, MNRAS, 394, 660

Chiaberge, M., Capetti, A., & Celotti, A. 2001, MNRAS, 324, L33

Cisternas, M., et al. 2011, ApJ, 726, 57

Cohen, M., Walker, R. G., Carter, B., Hammersley, P., Kidger, M., & Noguchi, K. 1999, AJ, 117, 1864

Colavita, M. M., & Wizinowich, P. L. 2003, in Society of Photo-Optical Instrumentation Engineers (SPIE) Conference Series, Vol. 4838, Society of Photo-Optical Instrumentation Engineers (SPIE) Conference Series, ed. W. A. Traub, 79–88

Cornwell, T. J. 2009, A&A, 500, 65

Bibliography

Courvoisier, T. J.-L. 1998, A&A Rev., 9, 1

Cox, A. N., ed. 2000, Allen's astrophysical quantities (New York: AIP Press; Springer)

Creech-Eakman, M. J., et al. 2010, in Society of Photo-Optical Instrumentation Engineers (SPIE) Conference Series, Vol. 7734, Society of Photo-Optical Instrumentation Engineers (SPIE) Conference Series

Croton, D. J., et al. 2006, MNRAS, 365, 11

Davies, R. I., Mueller Sánchez, F., Genzel, R., Tacconi, L. J., Hicks, E. K. S., Friedrich, S., & Sternberg, A. 2007, ApJ, 671, 1388

Di Matteo, T., Springel, V., & Hernquist, L. 2005, Nature, 433, 604

Díaz-Santos, T., Alonso-Herrero, A., Colina, L., Ryder, S. D., & Knapen, J. H. 2007, ApJ, 661, 149

Duchêne, G., & Duvert, G. 2007, New A Rev., 51, 650

Dyck, H. M. 2000, in Principles of Long Baseline Stellar Interferometry, ed. P. R. Lawson, 185–+

Elitzur, M., & Shlosman, I. 2006, ApJ, 648, L101

Espada, D., et al. 2009, ApJ, 695, 116

Evans, D. A., Kraft, R. P., Worrall, D. M., Hardcastle, M. J., Jones, C., Forman, W. R., & Murray, S. S. 2004, ApJ, 612, 786

Feain, I. J., et al. 2009, ApJ, 707, 114

Fienup, J. R. 1978, Optics Letters, 3, 27

Friedrich, S., Davies, R. I., Hicks, E. K. S., Engel, H., Müller-Sánchez, F., Genzel, R., & Tacconi, L. J. 2010, A&A, 519, A79+

Galliano, E., Alloin, D., Pantin, E., Lagage, P. O., & Marco, O. 2005, A&A, 438, 803

Gallo, L. C., Brandt, W. N., Costantini, E., Fabian, A. C., Iwasawa, K., & Papadakis, I. E. 2007, MNRAS, 377, 391

Gandhi, P., Horst, H., Smette, A., Hönig, S., Comastri, A., Gilli, R., Vignali, C., & Duschl, W. 2009, A&A, 502, 457

Gebhardt, K., Adams, J., Richstone, D., Lauer, T. R., Faber, S. M., Gültekin, K., Murphy, J., & Tremaine, S. 2011, ApJ, 729, 119

Glass, I. S. 1999, Handbook of Infrared Astronomy (Cambridge University Press)

Bibliography

Glindemann, A., et al. 2000a, in Society of Photo-Optical Instrumentation Engineers (SPIE) Conference Series, Vol. 4006, Society of Photo-Optical Instrumentation Engineers (SPIE) Conference Series, ed. P. Léna & A. Quirrenbach, 2–12

Glindemann, A., Hippler, S., Berkefeld, T., & Hackenberg, W. 2000b, Experimental Astronomy, 10, 5

Gorjian, V., Werner, M. W., Jarrett, T. H., Cole, D. M., & Ressler, M. E. 2004, ApJ, 605, 156

Greene, J. E., et al. 2010, ApJ, 723, 409

Guainazzi, M., Bianchi, S., Matt, G., Dadina, M., Kaastra, J., Malzac, J., & Risaliti, G. 2010, MNRAS, 406, 2013

Gültekin, K., et al. 2009, ApJ, 698, 198

Haas, M., Siebenmorgen, R., Pantin, E., Horst, H., Smette, A., Käufl, H.-U., Lagage, P.-O., & Chini, R. 2007, A&A, 473, 369

Hajian, A. R., et al. 1998, ApJ, 496, 484

Haniff, C. 2007, New Astronomy Review, 51, 565

Haniff, C. A., et al. 2000, in Society of Photo-Optical Instrumentation Engineers (SPIE) Conference Series, Vol. 4006, Society of Photo-Optical Instrumentation Engineers (SPIE) Conference Series, ed. P. Léna & A. Quirrenbach, 627–633

Hao, L., et al. 2005, ApJ, 625, L75

Hardcastle, M. J., Cheung, C. C., Feain, I. J., & Stawarz, L. 2009, MNRAS, 393, 1041

Hardcastle, M. J., Worrall, D. M., Birkinshaw, M., & Canosa, C. M. 2003a, MNRAS, 338, 176

Hardcastle, M. J., Worrall, D. M., Kraft, R. P., Forman, W. R., Jones, C., & Murray, S. S. 2003b, ApJ, 593, 169

Häring, N., & Rix, H.-W. 2004, ApJ, 604, L89

Harris, G. L. H., Rejkuba, M., & Harris, W. E. 2010, PASA, 27, 457

Hecht, E. 2001, Optics 4th edition (Addison Wesley)

Henkel, C., Peck, A. B., Tarchi, A., Nagar, N. M., Braatz, J. A., Castangia, P., & Moscadelli, L. 2005, A&A, 436, 75

Hennawi, J. F., et al. 2006, AJ, 131, 1

HESS Collaboration: F. Aharonian. 2009, ApJ, 695, L40

Ho, L. C. 2008, ARA&A, 46, 475

Högbom, J. A. 1974, A&AS, 15, 417

Hönig, S. F., Beckert, T., Ohnaka, K., & Weigelt, G. 2006, A&A, 452, 459

Horst, H., Duschl, W. J., Gandhi, P., & Smette, A. 2009, A&A, 495, 137

Horst, H., Gandhi, P., Smette, A., & Duschl, W. J. 2008, A&A, 479, 389

Israel, F. P. 1998, A&A Rev., 8, 237

Jaffe, W., et al. 2004, Nature, 429, 47

Jaffe, W. J. 2004, in Society of Photo-Optical Instrumentation Engineers (SPIE) Conference Series, Vol. 5491, Society of Photo-Optical Instrumentation Engineers (SPIE) Conference Series, ed. W. A. Traub, 715–+

Jahnke, K., & Maccio, A. 2010, ArXiv e-prints

Kaeufl, H. U., Bouchet, P., van Dijsseldonk, A., & Weilenmann, U. 1991, Experimental Astronomy, 2, 115

Kellermann, K. I., Sramek, R., Schmidt, M., Shaffer, D. B., & Green, R. 1989, AJ, 98, 1195

Kervella, P., & Garcia, P. J. V. 2007, New Astronomy Review, 51, 658

Kishimoto, M., Hönig, S. F., Beckert, T., & Weigelt, G. 2007, A&A, 476, 713

Kishimoto, M., Hönig, S. F., Tristram, K. R. W., & Weigelt, G. 2009, A&A, 493, L57

Köhler, R., & Jaffe, W. 2008, in The Power of Optical/IR Interferometry: Recent Scientific Results and 2nd Generation, ed. A. Richichi, F. Delplancke, F. Paresce & A. Chelli, 569–+

Komossa, S. 2008, in Revista Mexicana de Astronomia y Astrofisica Conference Series, Vol. 32, Revista Mexicana de Astronomia y Astrofisica Conference Series, 86–92

Kormendy, J., Cornell, M. E., Block, D. L., Knapen, J. H., & Allard, E. L. 2006, ApJ, 642, 765

Koshida, S., et al. 2009, ApJ, 700, L109

Krabbe, A., Böker, T., & Maiolino, R. 2001, ApJ, 557, 626

Kraft, R. P., et al. 2000, ApJ, 531, L9

Bibliography

Krolik, J. H., & Begelman, M. C. 1988, ApJ, 329, 702

Labeyrie, A. 1970, A&A, 6, 85

Landt, H., Bentz, M. C., Peterson, B. M., Elvis, M., Ward, M. J., Korista, K. T., & Karovska, M. 2011, MNRAS, L237+

Lawrence, A. 1991, MNRAS, 252, 586

Lehnert, M. D., et al. 2010, Nature, 467, 940

Leinert, C., et al. 2003, Ap&SS, 286, 73

Lindblad, P. O. 1999, A&A Rev., 9, 221

Lord, S. D. 1992, A New Software Tool for Computing Earth's Atmospheric Transmission of Near- and Far-Infrared Radiation, Tech. Rep. 103957, NASA

Lumsden, S. L., Heisler, C. A., Bailey, J. A., Hough, J. H., & Young, S. 2001, MNRAS, 327, 459

Lynden-Bell, D. 1969, Nature, 223, 690

Maiolino, R. 2008, New A Rev., 52, 339

Maiolino, R., Marconi, A., Salvati, M., Risaliti, G., Severgnini, P., Oliva, E., La Franca, F., & Vanzi, L. 2001, A&A, 365, 28

Maiolino, R., & Rieke, G. H. 1995, ApJ, 454, 95

Malin, D. F., Quinn, P. J., & Graham, J. A. 1983, ApJ, 272, L5

Marconi, A., Schreier, E. J., Koekemoer, A., Capetti, A., Axon, D., Macchetto, D., & Caon, N. 2000, ApJ, 528, 276

Markevitch, M., Gonzalez, A. H., Clowe, D., Vikhlinin, A., Forman, W., Jones, C., Murray, S., & Tucker, W. 2004, ApJ, 606, 819

Markowitz, A. 2009, ApJ, 698, 1740

Marquardt. 1963, SIAM Journal on Applied Mathematics, 11, 431

Mason, R., Wong, A., Geballe, T., Volk, K., Hayward, T., Dillman, M., Fisher, R. S., & Radomski, J. 2008, in Presented at the Society of Photo-Optical Instrumentation Engineers (SPIE) Conference, Vol. 7016, Society of Photo-Optical Instrumentation Engineers (SPIE) Conference Series

Matt, G., Bianchi, S., D'Ammando, F., & Martocchia, A. 2004, A&A, 421, 473

Mazzarella, J. M., & Boroson, T. A. 1993, ApJS, 85, 27

Bibliography

Meisenheimer, K. 2008, in Science with the VLT(I) in the ELT era

Meisenheimer, K., et al. 2007, A&A, 471, 453

Meisner, J. A., Tubbs, R. N., & Jaffe, W. J. 2004, in Society of Photo-Optical Instrumentation Engineers (SPIE) Conference Series, Vol. 5491, Society of Photo-Optical Instrumentation Engineers (SPIE) Conference Series, ed. W. A. Traub, 725–+

Michelson, A. A., & Pease, F. G. 1921, ApJ, 53, 249

Milgrom, M. 1983, ApJ, 270, 365

Minezaki, T., Yoshii, Y., Kobayashi, Y., Enya, K., Suganuma, M., Tomita, H., Aoki, T., & Peterson, B. A. 2004, ApJ, 600, L35

Moran, E. C., Barth, A. J., Kay, L. E., & Filippenko, A. V. 2000, ApJ, 540, L73

Morganti, R. 2010, PASA, 27, 463

Müller, A., et al. 2010a, in Society of Photo-Optical Instrumentation Engineers (SPIE) Conference Series, Vol. 7734, Society of Photo-Optical Instrumentation Engineers (SPIE) Conference Series

Müller, C. 2010, Master's thesis, Friedrich-Alexander-Universität Erlangen-Nürnberg

Müller, C., Kadler, M., Ojha, R., Boeck, M., & Wilms, J. 2010b, in COSPAR, Plenary Meeting, Vol. 38, 38th COSPAR Scientific Assembly, 2310–+

Mundell, C. G., Wrobel, J. M., Pedlar, A., & Gallimore, J. F. 2003, ApJ, 583, 192

Nagar, N. M., Oliva, E., Marconi, A., & Maiolino, R. 2002, A&A, 391, L21

Narayan, R., & Yi, I. 1994, ApJ, 428, L13

Nemmen, R. S., Bonatto, C., & Storchi-Bergmann, T. 2010, ApJ, 722, 281

Neugebauer, G., Graham, J. R., Soifer, B. T., & Matthews, K. 1990, AJ, 99, 1456

Neugebauer, G., Oke, J. B., Becklin, E. E., & Matthews, K. 1979, ApJ, 230, 79

Neumayer, N., Cappellari, M., Reunanen, J., Rix, H.-W., van der Werf, P. P., de Zeeuw, P. T., & Davies, R. I. 2007, ApJ, 671, 1329

Ohnaka, K., et al. 2009, A&A, 503, 183

Osterbrock, D. E. 1981, ApJ, 249, 462

Panessa, F., & Bassani, L. 2002, A&A, 394, 435

Pauls, T. A., Young, J. S., Cotton, W. D., & Monnier, J. D. 2005, PASP, 117, 1255

Bibliography

Peng, E. W., Ford, H. C., Freeman, K. C., & White, R. L. 2002, AJ, 124, 3144

Perlmutter, S., et al. 1999, ApJ, 517, 565

Peterson, B. M. 1993, PASP, 105, 247

Peterson, B. M., et al. 2004, ApJ, 613, 682

Pier, E. A., & Krolik, J. H. 1992, ApJ, 401, 99

Pott, J., Malkan, M. A., Elitzur, M., Ghez, A. M., Herbst, T. M., Schödel, R., & Woillez, J. 2010, ApJ, 715, 736

Prieto, M. A., Reunanen, J., Tristram, K. R. W., Neumayer, N., Fernandez-Ontiveros, J. A., Orienti, M., & Meisenheimer, K. 2010, MNRAS, 402, 724

Przygodda, F. 2004, PhD thesis, Max-Planck-Institut für Astronomie

Puech, F., & Gitton, P. B. 2006, Interface Control Document between VLTI and its Instruments, Technical Report, VLT-ICD-ESO-15000-1826, Issue 5.0, Tech. rep., ESO

Quillen, A. C., et al. 2008, MNRAS, 384, 1469

Quillen, A. C., Brookes, M. H., Keene, J., Stern, D., Lawrence, C. R., & Werner, M. W. 2006, ApJ, 645, 1092

Quirrenbach, A. 2000, in Principles of Long Baseline Stellar Interferometry, ed. P. R. Lawson, 71–+

Quirrenbach, A. 2001, ARA&A, 39, 353

Raban, D., Heijligers, B., Röttgering, H., Meisenheimer, K., Jaffe, W., Käufl, H. U., & Henning, T. 2008, A&A, 484, 341

Raban, D., Jaffe, W., Röttgering, H., Meisenheimer, K., & Tristram, K. R. W. 2009, MNRAS, 394, 1325

Radomski, J. T., et al. 2008, ApJ, 681, 141

Radomski, J. T., Piña, R. K., Packham, C., Telesco, C. M., De Buizer, J. M., Fisher, R. S., & Robinson, A. 2003, ApJ, 587, 117

Read, A. M., & Pietsch, W. 1998, A&A, 336, 855

Reunanen, J., Prieto, M. A., & Siebenmorgen, R. 2010, MNRAS, 402, 879

Riess, A. G., et al. 1998, AJ, 116, 1009

Riffel, R. A., Storchi-Bergmann, T., & McGregor, P. J. 2009, ApJ, 698, 1767

Risaliti, G., Elvis, M., Fabbiano, G., Baldi, A., Zezas, A., & Salvati, M. 2007, ApJ, 659, L111

Risaliti, G., Maiolino, R., & Salvati, M. 1999, ApJ, 522, 157

Rivinius, Th. 2011, Very Large Telescope Paranal Science Operations MIDI User Manual, Tech. rep., ESO

Robaina, A. R., et al. 2009, ApJ, 704, 324

Robberto, M., & Herbst, T. M. 1998, in Presented at the Society of Photo-Optical Instrumentation Engineers (SPIE) Conference, Vol. 3354, Society of Photo-Optical Instrumentation Engineers (SPIE) Conference Series, ed. A. M. Fowler, 711–719

Robson, E. I., et al. 1993, MNRAS, 262, 249

Rodríguez-Ardila, A., & Mazzalay, X. 2006, MNRAS, 367, L57

Rothschild, R. E., Markowitz, A., Rivers, E., Suchy, S., Pottschmidt, K., Kadler, M., Mueller, C., & Wilms, J. 2011, ArXiv e-prints

Saha, S. K. 1999, Indian Journal of Physics Section B, 73, 553

Sales, D. A., Pastoriza, M. G., & Riffel, R. 2010, ApJ, 725, 605

Sanders, D. B., Soifer, B. T., Elias, J. H., Madore, B. F., Matthews, K., Neugebauer, G., & Scoville, N. Z. 1988, ApJ, 325, 74

Sandqvist, A. 1999, A&A, 343, 367

Sandqvist, A., Joersaeter, S., & Lindblad, P. O. 1995, A&A, 295, 585

Sarazin, M., Melnick, J., Navarrete, J., & Lombardi, G. 2008, The Messenger, 132, 11

Schartmann, M., Meisenheimer, K., Camenzind, M., Wolf, S., & Henning, T. 2005, A&A, 437, 861

Schartmann, M., Meisenheimer, K., Camenzind, M., Wolf, S., Tristram, K. R. W., & Henning, T. 2008, A&A, 482, 67

Schartmann, M., Meisenheimer, K., Klahr, H., Camenzind, M., Wolf, S., & Henning, T. 2009, MNRAS, 393, 759

Scharwächter, J., Eckart, A., Pfalzner, S., Moultaka, J., Straubmeier, C., & Staguhn, J. G. 2003, A&A, 405, 959

Scharwächter, J., Eckart, A., Pfalzner, S., Saviane, I., & Zuther, J. 2007, A&A, 469, 913

Schinnerer, E., Eckart, A., & Tacconi, L. J. 1998, ApJ, 500, 147

Schmidt, M. 1963, Nature, 197, 1040

Schreier, E. J., et al. 1998, ApJ, 499, L143+

Schweitzer, M., et al. 2008, ApJ, 679, 101

Scoville, N. Z., et al. 2000, AJ, 119, 991

Shakura, N. I., & Syunyaev, R. A. 1973, A&A, 24, 337

Siebenmorgen, R., Krügel, E., & Spoon, H. W. W. 2004, A&A, 414, 123

Simpson, C. 1998, MNRAS, 297, L39

—. 2005, MNRAS, 360, 565

Skelton, R. E., Bell, E. F., & Somerville, R. S. 2009, ApJ, 699, L9

Soifer, B. T., Bock, J. J., Marsh, K., Neugebauer, G., Matthews, K., Egami, E., & Armus, L. 2003, AJ, 126, 143

Soldi, S., et al. 2008, A&A, 486, 411

Spergel, D. N., et al. 2007, ApJS, 170, 377

Spoon, H. W. W., Keane, J. V., Tielens, A. G. G. M., Lutz, D., Moorwood, A. F. M., & Laurent, O. 2002, A&A, 385, 1022

Springel, V., et al. 2005, Nature, 435, 629

Storchi-Bergmann, T., Mulchaey, J. S., & Wilson, A. S. 1992, ApJ, 395, L73

Struve, C., Oosterloo, T. A., Morganti, R., & Saripalli, L. 2010, A&A, 515, A67+

Suganuma, M., et al. 2006, ApJ, 639, 46

Surace, J. A., Sanders, D. B., Vacca, W. D., Veilleux, S., & Mazzarella, J. M. 1998, ApJ, 492, 116

Swain, M., et al. 2003, ApJ, 596, L163

Tallon-Bosc, I., et al. 2008, in Presented at the Society of Photo-Optical Instrumentation Engineers (SPIE) Conference, Vol. 7013, Society of Photo-Optical Instrumentation Engineers (SPIE) Conference Series

ten Brummelaar, T. A., Bagnuolo, W. G., McAlister, H. A., Ridgway, S. T., Sturmann, L., Sturmann, J., & Turner, N. H. 2000, in Society of Photo-Optical Instrumentation Engineers (SPIE) Conference Series, Vol. 4006, Society of Photo-Optical Instrumentation Engineers (SPIE) Conference Series, ed. P. Léna & A. Quirrenbach, 564–573

Teng, S. H., et al. 2009, ApJ, 691, 261

Tingay, S. J., et al. 1998, AJ, 115, 960

Tingay, S. J., & Lenc, E. 2009, AJ, 138, 808

Tingay, S. J., Preston, R. A., & Jauncey, D. L. 2001, AJ, 122, 1697

Townes, C. H. 2000, in Principles of Long Baseline Stellar Interferometry, ed. P. R. Lawson, 59–+

Tristram, K. R. W. 2007, PhD thesis, Max-Planck-Institut für Astronomie, Königstuhl 17, 69117 Heidelberg, Germany

Tristram, K. R. W., et al. 2007, A&A, 474, 837

—. 2009, A&A, 502, 67

Tristram, K. R. W., & Schartmann, M. 2011, A&A, 531, A99+

Türler, M., et al. 2006, A&A, 451, L1

Tuthill, P., et al. 2010, in Presented at the Society of Photo-Optical Instrumentation Engineers (SPIE) Conference, Vol. 7735, Society of Photo-Optical Instrumentation Engineers (SPIE) Conference Series

Tuthill, P. G., Monnier, J. D., Danchi, W. C., Wishnow, E. H., & Haniff, C. A. 2000, PASP, 112, 555

Ulrich, M.-H. 2000, A&A Rev., 10, 135

Urry, C. M., & Padovani, P. 1995, PASP, 107, 803

van Boekel, R. 2004, PhD thesis, FNWI: Sterrenkundig Instituut Anton Pannekoek, Postbus 19268, 1000 GG Amsterdam, The Netherlands

van Boekel, R., Min, M., Waters, L. B. F. M., de Koter, A., Dominik, C., van den Ancker, M. E., & Bouwman, J. 2005, A&A, 437, 189

van der Wolk, G., Barthel, P. D., Peletier, R. F., & Pel, J. W. 2010, A&A, 511, A64+

Veilleux, S., Kim, D., Sanders, D. B., Mazzarella, J. M., & Soifer, B. T. 1995, ApJS, 98, 171

Veron, P., Lindblad, P. O., Zuiderwijk, E. J., Veron, M. P., & Adam, G. 1980, A&A, 87, 245

Véron-Cetty, M.-P., & Véron, P. 2006, A&A, 455, 773

Vignali, C., & Comastri, A. 2002, A&A, 381, 834

Bibliography

Weaver, K. A., Yaqoob, T., Mushotzky, R. F., Nousek, J., Hayashi, I., & Koyama, K. 1997, ApJ, 474, 675

Weedman, D. W., et al. 2005, ApJ, 633, 706

Weigelt, G. P. 1977, Optics Communications, 21, 55

Weiß, A., Kovács, A., Güsten, R., Menten, K. M., Schuller, F., Siringo, G., & Kreysa, E. 2008, A&A, 490, 77

Whysong, D., & Antonucci, R. 2004, ApJ, 602, 116

Willott, C. J., et al. 2007, AJ, 134, 2435

Wilson, A. S. 1996, Vistas in Astronomy, 40, 63

Wright, E. L. 2006, PASP, 118, 1711

Young, S., Hough, J. H., Efstathiou, A., Wills, B. J., Bailey, J. A., Ward, M. J., & Axon, D. J. 1996, MNRAS, 281, 1206

i want morebooks!

Buy your books fast and straightforward online - at one of world's fastest growing online book stores! Environmentally sound due to Print-on-Demand technologies.

Buy your books online at
www.get-morebooks.com

Kaufen Sie Ihre Bücher schnell und unkompliziert online – auf einer der am schnellsten wachsenden Buchhandelsplattformen weltweit! Dank Print-On-Demand umwelt- und ressourcenschonend produziert.

Bücher schneller online kaufen
www.morebooks.de

VDM Verlagsservicegesellschaft mbH
Heinrich-Böcking-Str. 6-8
D - 66121 Saarbrücken

Telefon: +49 681 3720 174
Telefax: +49 681 3720 1749

info@vdm-vsg.de
www.vdm-vsg.de

Printed by Books on Demand GmbH, Norderstedt / Germany